AF661209

Forest Ecology in India
Colonial Maharashtra
1850–1950

Neena Ambre Rao

FOUNDATION®
BOOKS

Delhi • Bangalore • Mumbai • Kolkata • Chennai • Hyderabad

Published by
Cambridge University Press India Pvt. Ltd.
under the Foundation Books imprint
Cambridge House, 4381/4 Ansari Road, Daryaganj, **New Delhi** 110002

Cambridge University Press India Pvt. Ltd.

C-22, C-Block, Brigade M.M., K.R. Road, Jayanagar, **Bangalore** 560 070

Plot No. 80, Service Industries, Shirvane, Sector-1, Nerul,
Navi Mumbai 400 706

10 Raja Subodh Mullick Square, 2nd Floor, **Kolkata** 700 013

21/1 (New No. 49), 1st Floor, Model School Road, Thousand Lights,
Chennai 600 006

House No. 3-5-874/6/4, (Near Apollo Hospital), Hyderguda,
Hyderabad 500 029

© Cambridge University Press India Pvt. Ltd.
First Published 2008

ISBN 978-81-7596-549-2

All rights reserved. No reproduction of any part may take place without the written permission of Cambridge University Press India Pvt. Ltd., subject to statutory exception and to the provision of relevant collective licensing agreements.

Published by Manas Saikia for Cambridge University Press India Pvt. Ltd. and printed at Brijbasi Art Press Ltd.

Contents

Preface

Human life being closely dependent on the natural resource base for its survival, conservation and exploitation of natural resources have been an age-old phenomena. Consequently, every society has a natural resource use strategy that has evolved over thousands of years. The natural resource use strategy that evolved in India before the pre-colonial period had some distinct features specific to the region, the same being a derivative of factors such as the level of contemporary technology, population, demand pattern, political conditions and socio-cultural attitudes of the society towards nature. A number of foreign invasions and the consequent integration of migrant groups from other civilisations into Indian society as well as the changing ecological realities of the region influenced and altered these strategies time and again.

This book also traces strategies in the specific geographical region of Maharashtra, which was a part of the erstwhile Bombay Presidency. It further analyses the change in these strategies during the British colonial period and the impact on the people of this region, which gave rise to discontent in society. The colonial history is grounded in the socio-economic conditions of rural Maharashtra during the eighteenth century, particularly the nature of the relationship between rural communities and their living environment, the customary rights of the community, the usage pattern of natural resources such as forests and the cultural attitudes of people.

The study of evolution of colonial forest policies is the basic thrust of the book, especially British forest policies and their influence on the rural communities in Maharashtra. The ideological underpinnings that influenced colonial forest policies have been examined in detail. The environmental history of British India was closely interwoven with the bitter struggles organised by peasants and tribals to regain control of their living environment. The book also offers a brief story of the struggles of local communities against British colonial forest policies and their influence on the anti-colonial struggles in the early part of the twentieth century.

I wish to thank the Maharashtra State Archives, Bombay; Peshwa Daftar, Poona; Bharat *Itihas Samshodhak* Mandal, Poona; Deccan College, Poona; Servants of India Society Library, Poona; Andhra Pradesh State Archives, Hyderabad; Forest Departments of Poona and Hyderabad for their cooperation.

I thank Professor Atlury Murali, my supervisor, for his guidance and support. I also extend my gratitude to all those who helped me complete this monograph specially my colleagues Dr. Abdul Thaha (Hyderabad), Dr. Laxman Satya (Pennsylvania), and Suzanne Benally (Boulder, CO) for their constant support and encouragement.

Last but not the least, I thank my husband Suyodh for being a source of inspiration and my son Harshu for his loving presence in my life. Finally, I dedicate this work to my parents, Mr. Manohar Ambre and Mrs. Rajani Ambre. Without their inspiration and encouragement I could not have brought such a daunting task to fruition.

Neena Ambre Rao

Glossary

Gairan (Deccan) and *Gurucharan* (Konkan) – Land set apart for the free grazing of the village cattle.

Gaobhag – The portion of the forest area within the limits of any village set apart before the passing of the Forest Act to supply the local wants of the village, as distinguished from the imperial reserves.

Kharif – *Kharif* land in Konkan means land growing rice as used in the Bombay Forest Commission Report, 1887.

Knees – These are curved teak logs.

Kuran – Grass meadow yielding grass for fodder, unassessed and unoccupied but sold by yearly auction for grass and grazing or reserved for fodder by the government.

Parigh – *Parigh* is used to denote the area of unassessed waste, *varkas* land, wherever situated within the village boundaries which remains after all the land included in private holdings and all the land set apart for special purposes such as forest, grazing, village site etc. has been separately measured.

Varkas – This term is generally used to denote *kharif, rabbi,* garden or salt marsh land. It includes *Parigh.*

Malki number – Refers to any survey number in which the occupant has property rights. In practice it is now used in the Konkan to denote any survey number whether rice, *varkas* or garden land, held under the survey tenure.

Rab – This was a practice of burning layers of cow-dung tree-loppings, shrubs, leaves, grass and hay and various combinations in the fields before the commencement of rice cultivation.

Rabbi – Not used, as in the Deccan, to denote the late that is cold whether harvest but to distinguish from rice land to *varkas* land, a variety of alluvial formed dry crop land.

Shindad – This is that portion of *varkas* which is uncultivated and kept for the production of *rab* material.

Tahal – The loppings of the trees and shrubs used for *rab* cultivation were termed as *tahal*.

Vancharai – Refers to grazing.

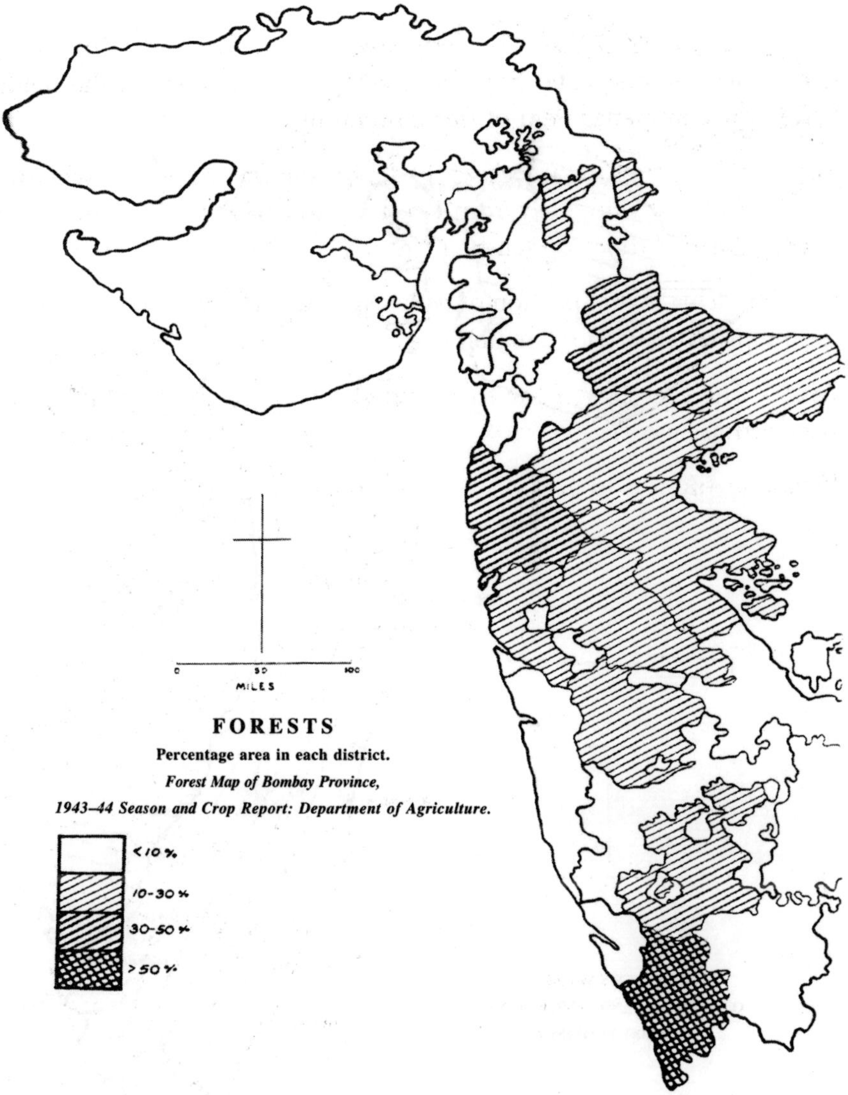

Source: Maharashtra State Archives, Bombay. Forest Map of Bombay Province.

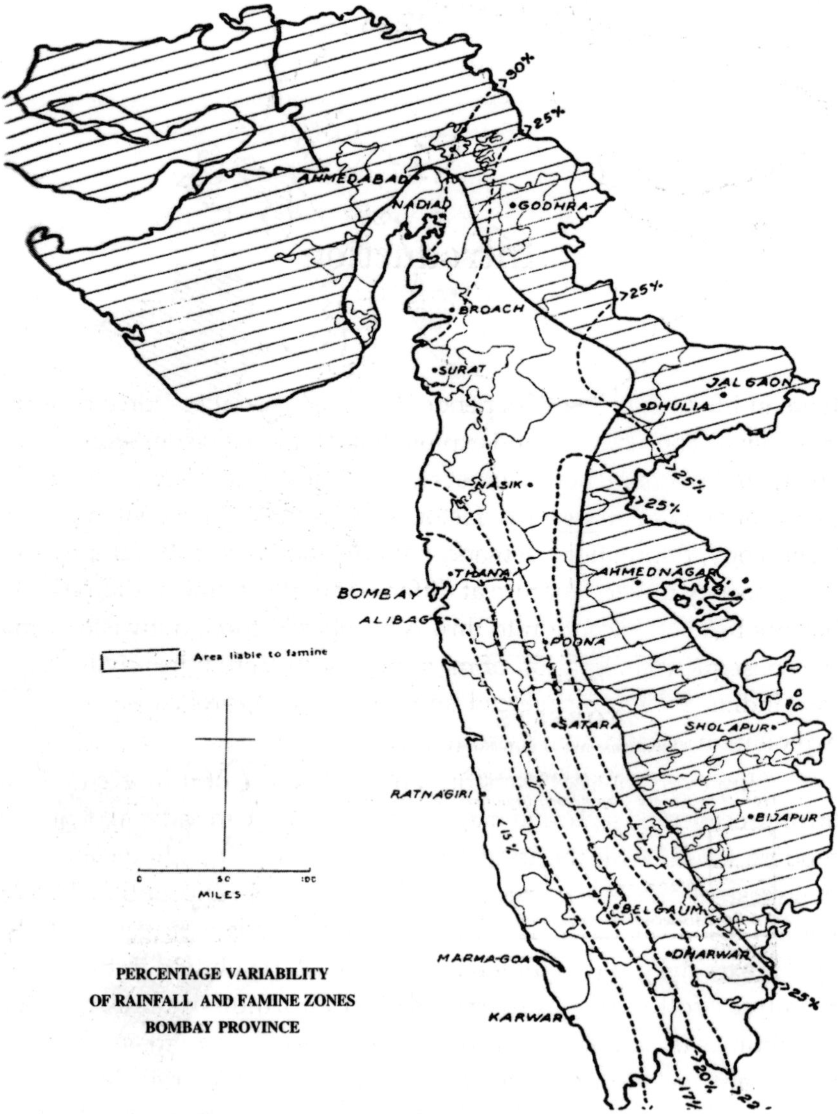

PERCENTAGE VARIABILITY OF RAINFALL AND FAMINE ZONES BOMBAY PROVINCE

Source: Climatological Tables: India Metereological Department, Government Photozinco Press, Poona, 1950.

1

Introduction

Human life being closely dependent on the natural resource base for its survival, conservation and exploitation of natural resources has been an age-old phenomenon. However, the quality and magnitude of this exploitation is a derivative of factors such as the level of contemporary technology, population, demand pattern, and socio-cultural attitudes of the society towards nature at different points of time. Indian society has not been an exception to this. A number of foreign invasions and the consequent integration of migrant groups from other civilisations into Indian society influenced and altered its approach towards the exploitation of its natural resource base.

However, the real transformation of the use of resource base on an unprecedented scale and magnitude began with the advent of British modern times. With the Industrial Revolution in their homeland, the main task of British administrators in India was to expropriate India's natural resources for their own imperial demands that were dictated by their military and commercial interests. A revolution in the use of resources following industrialisation enormously enlarged the possibilities of resource transformation from one form to another and their transportation over long distances. The Industrial Revolution created a zest among human beings, particularly from the West, to conquer nature. India being one of the richest colonies of the British in terms of its natural resources was an obvious target for the fulfillment of their rapacious greed. The legitimacy needed to appropriate the resources in the colonies was secured by passing several laws. The constitution of the Forest Department and the enactment of several Forest Acts were part of such attempts in colonial India.

The British systematically and legitimately exploited India's forests. In the process, they deprived India's rural population of the ecological base on which it was dependent for its livelihood since times immemorial. The process of commercialisation introduced by the British did not leave any fraction of Indian society untouched. The transformation from sustainable style of living to commercialisation was not very smooth for the rural population. The process of adjustment took a heavy toll causing resistance among the various sections of the society. Nevertheless, the system created by them, which initiated the process of ecological change, is continuing even after the end of British rule in India.

There are mainly two approaches to the study of forest conservation policies of British. Guha and Gadgil describe the pre-colonial period as the period of better ecological equilibrium. Their description gives an impression of a society that was insulated from any sort of external influence prior to the British rule.[1] The autonomous villages by virtue of their social institutions like the caste system managed their ecological affairs smoothly without disturbing the ecological equilibrium. It was only after the British onslaught on Indian society and its natural resources that the ecological equilibrium was disturbed badly.

In contrast to this, is the approach of Richard Grove. According to him, during the early phase of British forestry, personnel belonging to professions such as medicine and botany, who were genuinely concerned about the conservation of forests to maintain sound climatic conditions and prohibit phenomenon such as soil erosion, were given the charge of forests. It was they who initiated the process of scientific conservation of forests in India.

This book does not claim to be inclined towards any one of these approaches. Instead an attempt has been made to examine the commercial orientation given to the conservation of forests and its

1 Guha and Gadgil, *The Fissused land: An Ecological History of India*, OUP, New Delhi, 1992.

effects on Indian Society. While one does not deny the concern of some of the British officers regarding the effects of deforestation on India's climate, its connection with soil erosion and drought conditions, it cannot be denied that the most important motive was always the State's demand for timber, predominantly for the Royal Navy and other purposes of war .

It is also important to take into consideration the fact that though there was no ideal 'ecological equilibrium' in pre-British Maharashtra, awareness regarding the importance of preservation of trees was definitely in existence. It was ingrained into the people through their customs, traditions and religious attitudes. Their demand for natural resources was low because of their socio-cultural attitudes and the nature of material culture. Backwardness of technology was yet another important factor responsible.

Taking these into consideration, an attempt has been made in the present study to examine the British approach to forest conservation in Maharashtra and the socio-cultural background in which these efforts were made. Did commercial interests predominantly underlie their forest policies or was it a genuine concern for the depletion of forests? The way they responded to the demand for timber from their home country from time to time, and the way the forest policies evolved are the issues that have been examined in this book. The varying attitudes or positions taken by the different officers towards the issue of forestry, forest rights and community privileges have been highlighted in this work. An attempt has also been made here to examine the involvement of native intelligentsia in articulating the policies around the State or the European interests in India in general, and Maharashtra in particular.

The following questions have also been raised in the study: What were the socio-economic and cultural conditions and attitudes of local communities towards their natural resource base? Was the approach of native rulers towards natural resource management any different from those of the British? What was the approach of British towards the Indian forests? Was it solely from the commercial point of view that they began to preserve timber in Bombay Presidency? What was the background in which the formation of the Forest Department took place? What was the purpose behind the passing of the several Forest Acts? What were the different attitudes adopted by different Forest Officers and other bureaucrats of the British government about privileges and rights, which were to be granted to the people? Finally, what was the way in which the Forest Department was to cater to the needs of imperial demands for timber compared to the demands of the local population? What were the different reactions in different strata of the Indian society as a response to this? Was there any clash of ideologies, eruption of discontentment among the people?

Geographical Description of Maharashtra

The Maharashtra State, which is in the Western part of India along the Arabian Sea, is about three lakh square kilometres in area. It is the third largest state in the country. Gujarat, the Union Territories of Daman, Dadra and Nagar Haveli are to the northwest of Maharashtra, while Madhya Pradesh is to its north and east, Andhra Pradesh is to the south and east whereas Karnataka and Goa are to the south.

Maharashtra is also the third largest populated state in the country. It comprises of 31 districts, Mumbai suburban being the 31st district. For administrative purpose the State has been divided into six divisions: Konkan, Poona, Nasik, Aurangabad, Amaravati and Nagpur.

The physical features of Maharashtra divide the State into three distinct regions: Konkan, Ghatmatha and the plains of *Desh*. The narrow strip of land between the Arabian Sea and the Sahayadri Ranges is called Konkan. The Konkan coast stretches from 15.5 to 20.5 north latitude and spreads widthwise between 72.8 and 73.8 east longitude, the coastline from one end to another gradually receding by just 1 degree. Konkan comprises of five districts: Thana, Raigadh, Ratnagiri, Sindhudurg and Greater Bombay. Raigad district was earlier known as Kolaba in the Bombay Presidency. Sindhudurg comprises of a part of Ratnagiri district and an erstwhile principality of Sawantwadi. Greater Mumbai comprises of Suburban Bombay.

Throughout its length, Konkan is interspersed with numerous rivers and channels of sweet and salt water. Its mountainous tracts were once covered with thick forests. Cultivation of rice is possible mainly along the narrow margin of the coast where the soil is comparatively fertile. The mountain slopes with large patches of land carved out have been preserved for cultivation with great difficulty. Strangely this region receives plenty of rainfall but often suffers from scarcity of food due to its unproductive laterite soil. Marine products are therefore a vital source of nutrition for the people of this region.

The climate of the Konkan districts is humid and varies from one district to another. The southwest monsoon breaks with regularity in the days of June and ends in the middle of October. The rainfall is heavier on the slopes of the mountain ranges averaging between 90 and 100 inches. The rivers of Konkan are short in length due to the narrowness of the coastal strip. They flow from east to west, i.e. from the cliffs of the Sahayadris before joining the Arabian Sea with great speed. Ulhas, Savitri, Vashishthi and Shastri are the important rivers of Konkan.[2]

The Sahayadris are flat-topped mountains with their height varying from 20 metres on the coast to 2000 metres above sea level, from Parner Fort on Daman Ganga to Baba Budhan Hills in Mysore through Savantwadi and Goa. The Sahayadri range, running for about 750 km from north to south in Maharashtra, is part of the Western Ghats which separates the narrow coastal strip of Konkan from the plains known as *Desh*. The mountain range is the widest in Ulhas–Vaitarna valley in the Bombay region. There is lush green vegetation in Konkan all year due to a maritime climate and high humidity throughout the year. But the tops of hills are denuded in many places because of lack of soil. The extremely rocky regions such as Devgadh in Konkan look barren in dry months. The *Desh* looks green during and after the monsoon but gets parched after December and January. The high wall of Sahayadris serves as a barrier for the migration of species on either side. The mountain passes, locally called *ghat* or *bari*,

---

[2] K. R. Dixit, assisted by Charulata Patil, *Maharashtra in Maps*, published by Maharashtra State Board for Literature and Culture, Bombay, first edition, 1986, pp. 21–22; Dr. B. K. Apte, *A History of the Maratha Navy and Merchant Ships*, Maharashtra State Board for Literature and Culture, Bombay, 1972, pp. 11–15. Also see *Gazetteer of Bombay Presidency*, Vol. 13B, Thana, Vol. 13, Part I, 1882, pp. 1–4, *Imperial Gazetteer*, Kolaba–Janjira, Vol. 11, 1883, pp. 1–4, *Imperial Gazetteer*, Canara, Vol. 15, pp. 1–4.

are useful for traffic, trade and commerce. They form the routes going from *Desh* to Konkan and viceversa. There are 20–30 major passes, some of them are Bor Ghat between Bombay and Poona, Kasar Ghat between Bombay and Nasik, Pasarni Ghat between Wai and Mahabaleshwar, Phonda Ghat between Kolhapur and Goa, Varandha Ghat between Bhor and Mahad, Amba Ghat between Kolhapur, Ajra and Sawantwadi and South Konkan and Malshej Ghat between Junnar and Bombay. These ghats are generally narrow zigzag routes or paths with a scarp or canyon on one side and a valley on the other. They open huge vistas of scenery. Waterfalls, rivulets, rivers and cultivated fields are quite noticeable. Precipitation in the ghat sections and on mountain tops is always heavy. Dense vegetation develops in the valleys. The mountain tops are covered with thickest of stunted trees and bushes. In some places, near a source of river or a waterfall there are caves carved out in the brittle, multi-layered strata of the traps.

Hill forts on the peaks and spurs of Sahayadris have been known since the days of Kadambas. Principal among them are Sinhagarh, Raigarh, Pratapgarh, Purandhar, Panhala, Harishchandragarh, Shivneri, Doulatabad, Songarh, Aguada, and Phonda. The defence of the country depended upon the sturdiness of the hill forts as they were the real guardian sentinels of Maharashtra since the Bahamani rule. During the days of Shivaji, hill forts were maintained with great care for their immense strategic use.[3]

The Maharashtra Plateau

The region lying to the east of Sahayadris is called the Maharashtra Plateau. Traditionally, the term "Deccan" is applied to the whole tract of tableland stretching eastwards from the Satpura Ghats. The Maharashtra Plateau is a part of this Deccan Plateau. It includes many small pleateaux and river valleys. The area to the immediate south of the Satpura range along the Tapti Valley is at a lower elevation than

---

3 K. R. Dixit, assisted by Charulata Patil, *Maharashtra in Maps*, published by Maharashtra State Board for Literature and Culture, Bombay, First edition, 1986, pp. 22–23; *Gazetteer of India*, Maharashtra State, Botany and Flora of Maharashtra, Government of Maharashtra, Nagpur, 1987, pp. 1–54.

the rest and has its own climate and population. The Satmala Range separates this tract from the valley of Godavari and the Deccan begins from the south of these hills. The slope changes from west to the southeast; the elevation rises to 2000 feet at the edge of the ghats, overlooking the Konkan and falls gradually to about 1400 feet at the eastern limit. The rainfall is lighter than in Khandesh. Except in the west, it is far more uncertain than elsewhere in the state. Until the introduction of modern irrigation systems, the quality of soil further deteriorated and the dry open plain was scarcely remunerative. Owing to the frequent failure of the periodical rains, the tract to the east is known as a famine belt.

Nasik, Poona, Satara, Kolhapur and Ahmadnagar are the districts whose western side comprises of the Western Ghats. They are at a considerable altitude. The eastern side of these districts gradually merges into the Deccan plains. Hence, these districts comprise of the regions popularly known as *Desh* or Ghatmatha. By the side of the ghats, rivers and streams flow over labyrinthine paths, which open into low altitude basins of rivers in the valleys. The valleys are well-protected by two mountain spurs running parallel to each other and are fed by mountain streams, providing a unique climate and locations for plants to grow. These are known as *mavals*. There are 12 such *mavals* lying below Poona and another 12 below Junnar, where they are also called *nere*. This term is derived from the Persian word *nehar* which means a tributary or a small river or canal. Townships that developed here are named as Sangamner, Junner and Parner.[4]

The important river basins on the Maharashtra Plateau are as follows:

The Krishna Basin

The Krishna basin has its source at Mahabaleshwar in the Sahayadris. The Krishna flows across the southern part of the plateau. The important tributaries of this river are Koyna, Varna and Panchganga.

4 *Ibid*, *Imperial Gazetteer*, Poona, Vol. 18, 1885, pp. 1–10; Satara, 1887, Vol. 12, pp. 1–9.

The Bhima Basin

Bhima has its source at Bhima Shankar in the Sahayadris. The Indrayani, Mula, Mutha, Ghod, Nira and Sina are the main tributaries of the river Bhima. The Bhima basin is between the Mahadev ranges in the south and the Balghat plateau in the north.

The Godavari Basin

Godavari is the most important river in Maharashtra. It has its source at Triyambakeshwar in the Sahayadris in Nasik region. The Godavari basin is the largest river basin in Maharashtra. The Ajanta hills are to its north and the Balghat plateau to the south. Pravara, Sindphana, Manyad, Purna and Manjra are the important tributaries of Godavari.

The Penganga-Wardha-Vainganga Basin

Vainganga has its source in the Satpura ranges in Madhya Pradesh. Kanhan, Wardha and Penganga are its important tributaries. This basin is situated between the Ajanta range to the west, Darekasa and Chiroli hills in the east, Sironcha and Chandurgad hills in the south. The Penganga joins the Wardha and the Wardha joins the Vainganga. Their combined flow is called the Pranahita, which subsequently joins the Godavari.

The Tapti-Purna Basin

Tapti also has its source in the Satpura ranges, Purna being its main tributary. It flows from east to west. The river basin lies between the Satpura ranges in the north and Ajanta mountain ranges in the south.

The Narmada Basin

Narmada starts from Amarkantaka in Madhya Pradesh and flows through a small part of Maharashtra. Through Navapur it enters the alluvial plains of Gujarat and joins the Arabian Sea at Broach.[5] There are 41 tributaries of Narmada; to name a few are Hiran, Tendoni, Barna, Kolar, Man, Hatni, Banjar, Sher, Shakkar etc.

---

5 K. R. Dixit, assisted by Charulata Patil, *Maharashtra in Maps*, Maharashtra State Board for Literature and Culture, Bombay, First edition, 1986, pp. 24–25; *Gazetteer of India*, Maharashtra State, Botany and Flora of

The Climate of Maharashtra

The climate of Maharashtra is monsoonal. The Arabian Sea and the barrier of Sahayadri control the climate, the former by its cooling effect and latter by its altitude. The temperature does not remain the same throughout the year in Maharashtra. February, March, April and May are the hot months of the year. The temperature is relatively low from October to January. There is considerable variation in temperature, rainfall, humidity and topography in different regions of Maharashtra. For example, as we go higher from sea level, the temperature decreases and the air becomes cooler. The weather in the high ranges of the Sahayadris is cool even in summer, like at Mahabaleshwar in Satara district, Matheran in Raigadh district, Amboli in Sindhudurg district, Khandala-Lonavala in Poona district, Chikhaldara in Amaravati district and Toranmal in Dhule district. The Konkan strip being close to the sea, the climate is humid. It shows no extremes of either hot or cold. Summer is very hot in Vidarbha and like the Mediterranean, hot and humid in Khandesh. A peculiar type of climate prevails in the *mavals*. Their climate is very salubrious as is evident in cities like Poona, Nasik, Kolhapur, Belgaum and Satara. The average rainfall is not much and the altitude is also not high (365–610 metres approx.). Both are closer to the sea and not far from Sahayadris. Unlike Bombay or Konkan, the humidity too is neither low nor high.

The climate of Khandesh and Tapti valley is very hot in summer and cold in winter. They, however, do not face droughts like that in Sholapur, Ahmadnagar or Daund as the water-laden winds from Arabian Sea blow through them.

The Vidharbha has deep black cotton soils, which absorb a lot of heat during summer months from April to June and release it during the night. In fact, Akola is one of the hottest places in India, the highest temperature recorded being 120 °F. In eastern Vidarbha, at Chandrapur, Bhandara and Nagpur, the summers are dry, though the duration or

---

Maharashtra, *Gazetteer of India*, Maharashtra State, Botany and Flora of Maharashtra, (ed.), Chaudhari, Government Press, Nagpur, 1987, pp. 47; *Imperial Gazetteer of India*, Poona, Satara, *op.cit*, pp. 1–12, Nasik, Vol. 5, 1883, pp. 1–10, Khandesh, Vol. 12, 1880, pp. 1–11.

variation in temperature during the day time are not as high as in Akola. Pre-monsoon showers and thunderstorms too are not uncommon here.[6]

Rainfall Pattern

The main precipitation in Maharashtra comes during the monsoon. There is no snowfall anywhere, nor is mist seen. However, fog is there in the regions adjoining Belgaum and in the valleys of Khandala, Mahabaleshwar and Melghat during the pre- and post-monsoon periods, particularly early in the morning when the humidity is high. The variation in day and night temperatures is wide. There is no uniformity in rainfall over different parts of Maharashtra. Occasional hailstorms occur over the Deccan Plateau in April and May. The Konkan and coastal regions receive rains regularly averaging about 2540 mm per year. However, most of the precipitation flows rapidly to the sea due to the steep incline of the land, as its lateritic rocks and soil are poor in retaining water.[7] Hence, the land is far from fertile. Only flat patches in the small valleys or perched on hilltops are suitable for paddy cultivation. For 4 months of the rainy season there is incessant rain, followed by its conspicuous absence during the rest of the 8 months of the year. Still the humidity does not go below 60 per cent. On the *Desh* side, the rainfall is relatively high during these 4 months but the mean relative humidity is as high as 80 per cent in July and

---

6 *Gazetteer of India*, Maharashtra State, Botany and Flora of Maharashtra, Government of Maharashtra, Nagpur, 1987, pp. 16–17; K. R. Dixit, assisted by Charulata Patil, *Maharashtra in Maps*, Maharashtra State Board for Literature and Culture, Bombay, First edition, 1986, pp. 21–24.

7 The heavy rainfall in Konkan results in leaching of the rocks and their hardening. The laterites, a hydro-thermal product of withering, are derived from basalt and are rich in iron and aluminium. Due to its acidity, the lateritic cover does not support any vegetation. Only in the valleys and narrow riverine plains paddy is grown. On the valley side slopes particularly on the upper parts, the sub-lateritic horizons can be reached by trees. C. D. Deshpande, *Geography of Maharashtra*, National Book Trust, Delhi, pp. 92–95.

August and the lowest in March and April (varying from 25 to 30 per cent). On an average the highest annual rainfall in Vidharba is 1736 mm in Dhanora in Chandrapur. The lowest rainfall recorded is 448 mm at Akluj in Sholapur district. However, at Mahabaleshwar in the Sahayadris the average is 6226 mm. Hence, it is clear that the pattern of rainfall in Maharashtra is highly erratic. The highest number of rainy days is between 120 and 140 and the lowest from 25 to 40. The variation in the dates of the onset and cessation of the rainy season, the number of rainy days and the frequency and duration of dry spells, govern the character of the rainy season. Hence a large part of the State suffers from crop failures or periodical drought. In such circumstances, farming and cropping patterns have to be adjusted to the variations in the rainy season. Since this is not practically possible the variations in the pattern of rainfall cause natural calamities. Historically droughts and famines are not unfamiliar to Bombay Presidency, however in present times, the State administration is technologically advanced and economically sound to handle the problem in a better way by providing immediate famine relief measures, unlike the pre-Independence period when the only option people had was to migrate from one place to another.[8]

Forests of Maharashtra

A forest is a plant community naturally formed or both grown and mixed.[9] According to the statistics provided by the Forest Department of Maharashtra, the forests in the State occupy 66,725 square kilometres of area, amounting to around 21.8 per cent of the total area of Maharashtra. This figure, however, is deceptive as large part of the area identified as forests either has very sparse vegetal cover or is barren. Both Forest and Revenue Departments are guilty of showing exaggerated figures of areas under forests which are far from reality. There is no definition in terms of the volume of vegetal material or the

---

8 V. Subramanian, *Parched Earth: The Maharashtra Drought 1970–73*, Government of Maharashtra, 1975, pp. 23–31.

9 *Gazetteer of India*, Maharashtra State, Botany and Flora of Maharashtra, (ed.), K. K. Choudhary, Government of Maharashtra, 1987, p. 41.

frequency of trees per unit area to include a patch of land in the forest. What these departments provide is the legal status of the land use category as shown in records.[10]

According to the *Botony Gazetteer*, 11,192 hundred hectares are fallow lands and 13,799 hundred hectares are grazing lands. Area under agriculture is 68.8 per cent or 30.58 crore hectares.[11]

Forests in Maharashtra are mainly concentrated in the following regions due to their geographical location and terrain. These are:

1. Chandrapur-Bhandara at the extreme east of the State.
2. The valleys of Tapti in Khandesh and Satpura ranges, though under a different climatic set-up.
3. The western and eastern flanks of Sahayadris, mainly in the western ghats and few of the other forests in an edaphic climax such as mangroves, beach forests and forests on riverine terraces etc.
4. The spurs of Satpuras starting from the main ranges in Melghat extending eastwards on the *Desh* side which are only dry deciduous or scrub vegetation or forests with stunted growth on top of the ghats.

The best forests in Maharashtra are in the Chandrapur-Bhandara region and at a few places in the valleys of Sahayadris, both on the east and west side, apart from the Melghat and in low hills of Ajanta-Satnala ranges. Due to climatic peculiarities, there cannot be a specific system of classification of forests. In many places, the present day forests represent not the climatic climax vegetation but a degradation due to human interference, soil erosion and geophysical changes. For example, when large trees are auctioned in coupes in a forest, and timber and bamboo cut, the area quickly gets covered by *Lantana camara*. However, these are degrading changes and can be accepted

---

10 K. R. Dixit, assisted by Charulata Patil, *Maharashtra in Maps*, Maharashtra State Board for Literature and Culture, Bombay, First edition, p. 43.

11 *Gazetteer of India*, Maharashtra State, Botany and Flora of Maharashtra, (ed.), K. K. Choudhary, Government of Maharashtra, 1987, p. 41.

only under certain limitations and modification. Champion and Seth (1968) system is applied for the classification of forests in Maharashtra.[12]

Tropical Wet Evergreen Forests

The annual mean temperature in these forests is 27.2°C, rainfall 3292 mm and humidity 77 per cent. There is a canopy of multi-layered evergreen trees more than 45 metres high containing numerous epiphytes and a few climbers. In small pockets and valleys, they are found on the western face in the higher ranges of Sahayadris as on the western slopes of south Konkan in the Western Ghats[13], apart from the hilly areas of eastern Vidarbha. The trees include *Mesua ferrea, Dysoxylum malabaricum, Calophyllum inophyllum, Artocarpus* sp; *Hopea wightiana* and *Syzygium cumini*, mango, *sisson*, jackfruit, *jambul, hirda*, bamboo etc. These forests are thick with trees and creepers.[14]

Tropical Moist Deciduous Forests

The annual mean temperature in these forests is 27.1°C, rainfall 1324 mm and humidity 56 per cent. They occur throughout the moist part of Maharashtra receiving moderate rainfall. The top storey of vegetation consists of heavily buttressed, predominantly deciduous

---

12 *Ibid*, p. 46; Also see C. D. Deshpande, *Geography of Maharashtra*, National Book Trust, Delhi, pp. 68–79.

13 Such type is usually met with along the Western Ghats from Canara to Bombay. In this type of forest the ground vegetation is almost absent. *Ibid*, p. 68.

14 There is some ambiguity about the forests in the high rainfall areas of Western Ghats. These are classified by the Forest Department as 'sub-tropical broad-leaved hill forests'. One wonders if these could be grouped into tropical wet evergreen forests though the forests of Mahabaleshwar, Matheran, Bhimashankar and Amboli Ghat have nothing of the luxuriance of the tropical wet evergreen forests. Their stunted growth is owing to edaphic constraint. The latiritic crust that has developed at Mahabaleshwar and Matheran is detrimental to the growth of trees. K. R. Dixit, *Maharashtra in Maps*, assisted by Charulata Patil, *Maharashtra in Maps*, Maharashtra State Board for Literature and Culture, Bombay, First edition, 1986, pp. 44–45.

species and irregular trees 40 metres or more in height. The second storey is of predominantly deciduous plants mixed with some evergreens, the undergrowth shrubby and with dense climbers. In places cane and bamboo grow as undergrowth. This form is mainly seen in Allapalli forests of Chandrapur. The trees comprise of *Tectona grandis, Semecarpus anacardium, Salmalia malabarica, Lagerstroemia parviflora*. Locally the trees grown here are called teak, *khair, banian, peepul, neem, ain, al, palas,* tamarind, *amla* etc. These forests are not as dense as the evergreen forests.[15] Besides the large tracts of Chandrapur, they are also found on the eastern slopes of the Sahayadris and also in Konkan, Melghat, Satpura, Satnala and Ajanta ranges.

Littoral and Swamp Forests

The annual mean temperature in these forests is 27.9°C, rainfall 1042 mm and annual humidity 73 per cent. They consist of evergreen species in variable density and height near the seacoast, estuaries and river deltas. The temperature remains constant and rainfall occurs in abundance. The humidity is high throughout the year. Mangroves of Raigad, Thana, Greater Bombay, Ratnagiri, Chiplur and Terekhol belong to this category.

The littoral forests represent an edaphic climax. They contain some small evergreen and deciduous trees. Surface creepers such as *Ipomoea biloba* and maritime grass *Spinifex littoreus* are abundant in areas where fresh water is available. Tall grass phragmites occur at some places on Malvan and Vengurla coast.

The mangrove scrub and strand plants form dense growth on muddy coastal flats and river deltas. *Sonneratia apetala, Aegiceras majus, Avicennia officinalis, Acanthus ilicifolius* grow mixed with creepers of

---

[15] These are the most representative and general type of vegetation, which prevail throughout the Sahayadris except in areas having evergreen vegetation. Teak is naturally the most representative and commercially the most important tree. In 1985, teak forests occupied around 11,500 square kilometres area which was about one-fifth of the area under Forest Department. C. D. Deshpande, *Geography of Maharashtra*, National Book Trust, Delhi, p. 70.

Tylophora indica. Behind these are the tropical riverine forests fringing the banks of the river at Janjira (Murud) etc.[16]

Southern Tropical Dry Deciduous Forests

The annual mean temperature here is 27.3°C and rainfall 825 mm. These forests consist of a number of deciduous species that dry up in summer months. Their upper canopy is not close and the trees are generally 20 metres high. Different species predominate in different areas, but they are not gregarious and heterogeneous. The canopy being made up of deciduous small trees or tall shrubs, the sunlight reaches the forest floor enabling good growth of grass with few climbers. Some like *Entada pursaetha* are large and woody. Bamboos are present. A few epiphytic ferns and orchids too occur. This type is abundant throughout the western region of Maharashtra and other parts of the Western Ghats.

The Southern Tropical Dry Deciduous Forests comprise of:

1. *Tall Teak Forests*: Here dry deciduous species such as *Terminalia tomentosa, Adina cordifolia* and *Schleichera trijuga* predominate with *Tectona grandis* being the main species.
2. *Southern Moist Mixed Deciduous Forests*: These are common in ghat sections and the Chandrapur area of Maharashtra. *Terminalia belerica, Pterocarpus marsupium, Salmalia malabarica, Anogeissus latifolia, Dalbergia sissoo* and *Dalbergia latifolia* occur in plenty.
3. *Southern Dry Mixed Deciduous Forests*: These comprise of *Boswellia serrata,* bamboos, *Bauhinia vahlii, Acacia torta* and *Acacia leucophloea.* Rainfall is 975–1500 mm. They occur on well-drained hillsides and are very common in Bhandara district. Also there are *Terminalia tomentosa, Lagerstroemia parviflora, Soymida febrifusa, Anogeissus latifolia, Acacia catechu, Emblica officinalis, Zizyphus xylophones,* etc. found. The forests

[16] In Maharashtra littoral forests are not as significant as in the delta areas of Bengal, AP and Orissa. These trees are used for fuel, and are also useful in preventing coastal erosion. K. R. Dixit, assisted by Charulata Patil, *Maharashtra in Maps*, Maharashtra State Board for Literature and Culture, Bombay, First edition, p. 46.

near Chaturshringi at Poona and Katraj belong to this type. However, due to human interference, the Chatursringi forests have got destroyed now.

4. *Dry Grassland*: Here *Sehima nervosum, Heteropogon contortus, Dichanthium annulatum* and *Themeda triandra* are in abundance on hills near Kalyan. These grasslands often deteriorate. The leafy fodder grasses get replaced by annual woody ones especially by *Heteropogon contortus, Aristida hystrix, Eragrostis aspera* and *Cymbopogon martini*. Various edific climax type of dry deciduous forests are known in Maharashtra such as babul forests between Daund and Poona, *Hardwickia binata* forests between Malegaon and Dhule and bamboo brakes in Melghat etc. They indicate secondary succession.
5. *Babhul or Babul Forests*: *Acacia arabica* grows gregariously in these forests. Undergrowth is very little and grass cover is thin for grazing. The original type has got considerably altered due to grazing. A similar type with *Balanites roxburghii, Acacia leucophloea, Prosopis spicigera, Azadirachta indica, Casearia grandis* and *Butea monosperma* form a similar degraded forest series from dry deciduous forests in Vidarbha, Marathwada hills, Kinwat etc. Human interference has affected these forests.

Tropical Thorn Forests

These forests have an annual mean temperature of 27 °C, rainfall 691 mm and humidity 47 per cent. They are low, open xerophytic forests containing thorny leguminous plants. *Acacia torta, Azadirachta indica, Phoenix sylvestris,* and *Butea monosperma* (palas) are commonly found in these forests. Around one-sixth of the land classified as forests comes under this category. Small patches of these forests are privately owned.[17]

Tropical Dry Forests

The average mean temperature in these forests is 27.9 °C, rainfall 1052 mm and humidity 74 per cent. These are low forest – 12 metre in height.

---

[17] *Ibid*.

They are restricted to a small coastal area adjoining Savantwadi in Maharashtra.

West Coast Semi-Evergreen Forests

They occur between wet evergreen and moist deciduous forests in small narrow strips in the Western Ghats.[18] Biotic factors have considerably altered the forests. Species found here are terminally paniculata, *Diospyros assimilis, Alysicarpus* etc.

All the aforementioned forests form a useful vegetal cover in Maharashtra, though only some of them are economically important. For example, the dry deciduous teak forests are the most important as they are open grasslands.[19]

Today the forests are owned, protected and exploited by different agencies. Out of the total area of forests in the State, 90 per cent is administered by the Forest Department. About 1.5 per cent is under the control of the Forest Development Corporation and another 7 per cent is under the supervision of the Revenue Department. The rest of the forests are in private ownership.

Sources

The sources utilised for the study are numerous and varied. They are inclusive of the records from Peshwa Daftar, which is a collection of papers since 1729 when the centre of Maratha power shifted from Satara to Poona. Written in Modi script, these papers contain correspondence not only political in nature but also of revenue, defence, police, prisons, judiciary and such varied aspects of government administration. The

---

[18] These are found in the regions of 80 inches to 100 inches of rainfall, where the moisture is not adequate to promote a true evergreen type of forest and yet is capable of supporting a semi-evergreen formation. They exist on the flank of Sahayadris. They are usually associated with Bamboo brakes. C. D. Deshpande, *Geography of Maharashtra*, National Book Trust, Delhi, pp. 70.

[19] *Gazetteer of India*, Maharashtra State, Botany and Flora of Maharashtra, (ed.), K. K. Choudhary, Government of Maharashtra, 1987, pp. 50–51.

papers have been numbered and preserved in different *rumals*. Each rumal contains a number of bundles called *Pudake*.

Along with this another important Marathi source available in printed form is *Wad Dairies*. They are a collection of correspondence regarding the Peshwas' administration that covers various aspects such as judiciary, crime, agriculture, forests etc.

The *English Records on Shivaji* contains 729 closely printed pages which include 1000 letters and 11 extracts from histories and contemporary narratives. It contains correspondence among the East India Company officers working in different factories in India regarding Shivaji and his territory. They highlight the importance of teak reserves of Konkan for the Company. Besides this, there are various important Marathi sources such as Ajnapatre, Shiv Charit Sahitya Khand etc.

Most of the English government sources were obtained from the Maharashtra State Archives, Mumbai, which contain various administrative reports, forest working plans, statistics of annual returns, quinquennial reports, Revenue Department files etc. These provide the necessary data on the basis of which it becomes possible to examine various aspects of British forest administration like the commercial orientation of forestry. Along with these there are reports of various committees formed to look into the grievances of the people related to the issue of forests. These reports give an idea of the nature and extent of people's discontentment due to various forest restrictions and also the attitude of the administration towards these grievances.

Of these reports, the *Bombay Forest Commission Report* in four volumes is the most voluminous. It contains an important collection of testimonies of 532 villagers from Kolaba and Thana districts belonging to different strata of society.

A small section on *vancharai* i.e. grazing during 1830s and 1840s could be traced in the records of the Alienation office at Peshwa Daftar. These records provide us the correspondence between the Collectors and the Revenue Commissioners of the respective divisions over the issue of grazing, common village pastures etc. Also, there is a discussion about felling of trees and killing of wild animals. The correspondence is available in manuscripts.

The district gazetteers, census reports etc, provide the basic data regarding the geography and history of the territory. Sources like the reports of the Sarvajanik Sabha, their quarterly journal, and newspapers like *Kesari* give us an idea of the extent of involvement of the contemporary intelligentsia in the issue of forestry and their efforts to bring forward people's problems.

Different issues of the *Indian Forester* belonging to the later nineteenth century represent the opinions of Forest Officers, who were actually involved in the formation and evolution of the forest policy.

2

Pre-Colonial Maharashtra

Recent concern for sustainable development has made it necessary for us to examine the consequences of our actions upon the environment. The overuse of natural resources has become a major international concern. The focus of this unit is the land distribution and forest management practices of the Maharashtra that were under Maratha control, namely the *Maval* region (part of Western Ghats), the high and arid plains of the *Desh* region, and the Konkan costal area. Marathas, beginning with the Bhosle dynasty, and later, the Peshwas, ruled this region from 1640 to 1818.

This type of study must consider the geographical diversity, and various political, socio-economic, cultural and religious influences present in modern-day India. In order to understand a society's relationship with nature, one must first analyse the socio-political and economic conditions that existed within its borders.

As we reconstruct the environmental history of this region, it is important to note that the area is comprised of three distinct geographical zones: the high, arid plains; the mountainous region and the wet, coastal areas. Hence, each area differs in terms of its available resources, farming practices, political history, and its interactions with outsiders. However, one can also see significant similarities between the populations of each geographical area, i.e. they share comparable socio-economic systems and common religious practices. These variables will serve as recurring themes in the pre-colonial history of this region.

Village Structure and Distribution of Land

The village was a self-supporting unit, whose patterns of agriculture were largely determined by its specific requirements. Since most agricultural production catered to the local needs, the activities of a given area were directed towards fulfilling the individual needs.

The economic life of the village was based predominantly upon subsistence farming, and nature was an integral part of civic life. Towns and commercial centres were comparatively rare. A town with a public market was known as a *kasba,* whose permanent population comprised of mainly artisans. Weekly *bazaars* were a special feature of the *kasba,* and often attracted people from surrounding villages. Thus the *kasba* was a place of considerable economic activity, and its shopping area or market was referred to as a *peth.* A town with a port was known as a *mire.*

The villages of *Desh* were surrounded by walls, which formed the border that was a line of defence against attack. Such walls can still be found in some modern villages, though in a dilapidated condition. The main entrance to the village was also a typical feature of these communities.

Village land was divided into two main parts: *kali,* or arable land, and *pandhari,* which was used for habitation. *Kali* included cultivated land, fallow ground, and wasteland. The common lands in a village were also part of *kali.*

In the case of villages in Konkan and *Maval,* the land was not divided into *kali* and *pandhari;* instead it was the land used in rice cultivation that was most prized. Less fertile land was known as *varkas zamin.* Crops such as millet were grown on such land. There were also large tracts of forest (*raan*), pastures or grazing lands (*gairan*), bodies of fresh water, coastal and marine areas, and to some extent, agricultural lands. These lands were controlled and managed by village institutions. For instance, *gairan* fell into the category of village commons meant for grazing. Then there were the village forests (*raan* or *kurans*) in which people inherited rights to collect wood by paying a nominal tax. There were also sacred groves (*deorai*).

The village allotted newcomers unoccupied parcels of land for building houses. The outlying areas of the village were called by a number of names, including *majra, wadi, padi* and *khari*.

The villages in Konkan were generally spread along the sea. These villages typically extended in a narrow strip about 2 miles long, with the sea on one side, and hills or fields on the other. A traditional house in a Konkan village had open space on all four sides, with an *aangan*, or porch in front, and a *paras*, in the back that was generally used for gardening and growing medicinal plants. The houses in a typical *Desh* village stood adjacent to one another without any space in between. The roofs of *Desh* houses were flat, in contrast to the sloping roofs of Konkan houses.

The village was a self-sufficient unit, and contained all the constituent elements of a miniature state. The needs of the village determined its industries and the division of labour. Its leading officials (whose positions were hereditary) were entitled to land and special perquisites from the state. They were known as *deshaks*. They were permanent residents of the village and served as administrative and judicial functionaries. Sumit Guha writes that "these officials were a kind of hybrid figure partly entrepreneur, partly official and partly territorial magnates."[1] They executed the orders of Maratha rulers with regard to the clearing and cultivation of new land, exemptions of certain persons from tax revenues and so forth.

A wide range of political and economic factors could influence the relationships between the ruler and hereditary officials. During the rule of Bajirao II (1789–1818) transfer of land in these offices frequently changed hands.

The majority of villagers paid land revenue and were called *mirasdars*, an Arabic term meaning 'hereditary owners'; other villagers, who did not own the land they occupied, were referred to as *uparis*, or strangers. Elphinstone and Chaplin note that when the British took control of this territory in 1818, *mirasdars* outnumbered *uparis*. By 1838, Skyes reports that 40–50% of Deccan inhabitants were registered

---

1 Deshpande, C.D., *Geography of Maharashtra*, Series: India–Land & People, Sr. Pub, Verry, Lawrence Incorporated, 1973, p. 21.

landholders, with another 20–25% serving as tenants and labourers. About 75% of the population was engaged in agriculture. Villages which experienced labour shortages would often invite *uparis* from other areas to farm the land. In due course, many of these *uparis* acquired proprietary rights over the land they worked.

Another group of inhabitants, known as *balutedars*, or artisans, provided a range of services to the local population. One well-known *balutedar* named Mahar, occupied a position of great prominence in his village, despite being an outcast. As the village watchman, he also reflected his values and concerns. His duties included preventing the unauthorised theft of firewood from village forests, and harvesting and delivering of wood to village households etc.

Other *balutedars* worked as carpenters or blacksmiths who manufactured goods for local use, with the surplus being sold to other villages. As payment for their services, these *balutedars* were given an annual allotment of food grown in the village, as well as a hereditary right to a tract of land for their service to the local government. The number of *balutedars* varied from village to village, and some eventually elevated themselves from *upari balutedars* to *mirasdar balutedars* through the acquisition of land.

The many tribes which inhabit this region can be divided into two basic categories: forest tribes and nomadic tribes. Among these were the *Koli*, *Varli*, *Katkari*, and *Thakur* groups who lived mainly in the forests, particularly in the areas of Nasik, northern Konkan, eastern Ratnagiri, and western Satara and Kolhapur. These groups were largely isolated from mainstream society and were intensely insular. Though there were occasional intertribal tensions between forest and nomadic tribes, they were able to co-exist with one another. The flexible nature of land ownership helped avert conflict between disparate interests.

However, villagers and nomads often had mutually beneficial economic relationships. For instance, *vaidhus* were a class of nomadic merchants who sold medicine and spiritual cures, and frequently set up shops in cloth tents on the outskirts of town. They referred to themselves as *mandur vaid* (doctors who sell medicine) or *pariksha vaid*, which translates as 'pulse-feeling doctor'. Their medicines were made from herbs and roots gathered from the forests.

Another group, known as *kolis*, were feared warriors whose resistance to British rule made them a thorn in the eye of colonial authorities. The Indian method of dealing with *kolis* was to first subdue them, and then install them as military commanders over the areas in which they resided. In the seventeenth century, when the Mughal emperor Shahjahan tried to impose land taxes upon recently conquered territories, he faced staunch opposition from *kolis*, who refused to let their estates be surveyed or taxed. This attitude persisted until recently, as the *kolis* argued that their land was a sovereign territory, that did not fall under the dominion of the central government.

Nomadic tribes such as the *Lamanis, Wanjaris, Kaikadis, Vaidus,* and *Bhamtas,* traded in grain and salt. They also tended various herd animals, such as cattle, buffaloes, sheep and occasionally camels. During the rainy months, they retired to the forests until the weather cleared.

The eastern parts of Poona and Satara districts formed the pastoral regions of the Maratha territory. Rainfall has always been very light in this region and agriculture not very productive due to the poor soil. Historically, this region was the connecting link between the trading centres of the west and agrarian regions of the east. Until the spread of irrigation, the economy of this region was based mainly on pastoralism and supported only a scant population.

Dhangars formed the main pastoral community, and during the fair weather months, they travelled westward toward the hills, going even so far as the Konkan, supplying manure to the people of the villages, for which they were paid in kind. So highly prized was their manure, that for one night, a flock of 20 sheep might produce enough manure to earn their owner 12–16 pounds (3 to 4 *adhelis*) of rice. Pen, Alibag, Mahad and other larger towns featured butchers known as *khatiks*, who did not keep animals. Instead, they bought one or two at a time, according to the demand.

Before the end of summer, these sheepherders moved back east, through the pastoral regions, because during the rainy season, the dampness of the hills could be harmful to the sheep. They lived in huts with walls of wattle reeds and roofs thatched with straw. While wandering, they raised temporary huts made of twigs and wild plantain leaves. The chief enterprise of *dhangars* was sheep and cattle breeding and blanket weaving.

Traditional Management of Land and Forests

While the nation's ruler was supposedly the owner of all land, those who cultivated it gradually came to acquire ownership. Thus did peasant proprietors – such as the *mirasdars* – who held most of the village land.

Arable land belonging to the government was usually given to *uparis* for a certain period. This in turn permitted the government to collect land revenue. Village land was sometimes divided into *khalisati*, whose revenue was paid to the ruler's treasury and *inamati*, whose taxes could be appropriated by various *inamdars* in lieu of direct payment from the state.

As mentioned earlier, land held by virtue of hereditary office was called *watan*. It appears that hereditary officials held around 20% of the land under cultivation. *Inamdars* were not required to pay land revenue like other inhabitants of the village; however, this did not in any way affect their rights over their land. Among the *inamdars* were a number of institutional landowners, such as religious institutions, which employed individual tenants to work on the land. Taxes from a given piece of land were sometimes directed toward the maintenance and upkeep of a religious site. Such lands were not transferable, either by the village or an individual. For example, Ramdas (a seventeenth century Marathi saint held in high reverence) had received a substantial estate from Shivaji for its maintenance at Chaphal.

Land was also held as *agrahara inam*, or land grant. For example, Shivaji once gave an entire village in Poona as a grant to a Brahmin. A special class of such grants were occasionally given, when the cultivators allowed a village artisan or labourer to cultivate a piece of reclaimed wasteland without paying any land tax. This was done in exchange for that labourer or artisan's share of the village produce.

Land described as *Tikanati Begari* are lands given to certain people in a village for rendering certain services without remuneration. These lands could be mortgaged but not sold, and could be taken back by the village for failure to provide the service. This type of land grant was common in the hilly regions of the Western Ghats.

As the ruler was the ultimate owner, land was reverted to him when no one had a title to it. The cultivators as well as the *inamdars* could sell their title to the land or their title to the land revenue. They could secure loans using their titles as collateral. For example, a letter from the *Wad Diaries*[2] indicates a sale in which 250 rupees were paid for eighteen acres of land with wells and trees.

Village officials could sell their right to the collection of land revenue only with the consent of the villagers. This ensured that those holding power made decisions that were acceptable to their constituents. There were instances where villagers took over the management of land revenue (with the approval of the government) because they were dissatisfied with the decisions of their local officials. The following letter from *Wad Diaries* is one such example:

> "The Khoti Watan of Nandvi Dhabole in Tarf Govele and C belong to government. Sheik Mohammed was appointed the agent of the government to carry on the duties of the Watan. The ryot being dissatisfied with the agent, offered to pay to government the amount of the farm (agreed upon by the agent) together with the profit of the Watan, if the management were entrusted upon them instead. The offer was accepted."[3]

Despite the high demand for arable land, there were many factors which impeded its sale and purchase. It was often difficult to find a buyer, particularly for the *'inam'* (grant) land. While several exchanges of land are found in records from the period, the general belief was that land held by virtue of one's hereditary office should not be bought or sold. Prahlad Niraji, the *nyayadhish* (judicial official) of Shivaji, stated in one of his verdicts:

---

2 Rao Bahadur Ganeshji Wad, selection from Satara, Raja and Peshma Diasies, Vol. 7, Letter No. 433, 1770–71, pp. 119, printed and published with the permission of the Government of Bombay by D.B. Phandis, 1911.

3 Ibid, Vol. 3, Letter No. 405, 1740–41, pp. 252.

> "Nobody should purchase the Watan, no Watandar should sell it, those who purchase it, should be punished and fined. The Watandar selling his Watan should be warned strongly."[4]

The purpose of this policy was to maintain the dignity of the *watan* and prevent it from becoming an ordinary commodity, which could be bought and sold indiscriminately. *Watans* retained their respected status, regardless of political changes. However, the selling of titles became an increasingly common occurrence, dating from the end of the eighteenth century.

The percentage of land under cultivation remained rather low throughout this period. Farmers had to labour hard to make even a subsistence living. During this time, Shivaji made anxious inquiries about the human resource available in Konkan villages in 1676. Because labour was a central requirement of agriculture, rulers extended incentives to those who were willing to bring fallow land under cultivation. Farmers faced difficulties such as unpredictable amount of rainfall, famines and frequent warfare, and these factors sometimes led to the wholesale migration of villages from one location to another. Oppressive taxation also contributed to such radical population shifts. In cases of stable rule, leaders made efforts to resettle villages and to find new villages. There were instances of rulers seizing the harvest of a particular village until its land revenues were paid. These same rulers often resorted to land seizure and the imprisonment of peasants to ensure compliance. At times, peasants in other villages were made to pay taxes owed by the inhabitants of another village. However, they never forfeited their *miras* because of an inability to pay land revenues.

Due to armed conflict, it was incumbent upon administrative officials to meet agricultural production goals. As warfare became more widespread, hereditary officials were seen as being responsible for the suffering of the people they governed. The purpose of the *inam* grants given to hereditary officials was to promote increased agricultural

---

4 A.R. Kulkarni, Shivkalin, Maharashtra (Marathi), Rajhans Publications, Poona, 1991, pp. 21–46.

production. To this end, village officials were authorised to offer incentives including advances for purchase of seeds, farming implements, and temporary exemption from land revenue. But if conditions became such that farmers began leaving villages, State officials were also seen as being culpable. Hence, as Fukozawa writes, "if the Maratha land revenue was heavy, its administration was efficient except under the last Peshwa and the government was open to complaints and ready to redress villagers' grievances."[5]

The government sometimes waived taxes during periods of famine or other natural disasters and gave *takavi*, or short-term loans, at rates below those charged by moneylenders. What's more, substantial amounts of government revenue were paid out within the village. According to Skyes, in a typical nineteenth century Deccan village, Rs. 3,022 (out of Rs. 8,522 collected) were paid out within the village.[6]

Revenue farming was one of the organisational means of agrarian restoration and expansion, which existed in the Deccan prior to Maratha rule. However, it became a common practice in the Maratha region, especially after 1761. Revenue farming eventually became the means by which rulers weathered financial crises. Baji Rao II (1790–1818) resorted to this system of farming by auctioning land to the highest bidder. This system eventually spread to all levels of administration. There was no redress for the peasant, consequently many gave up farming.[7]

Although the area of land under cultivation increased during Peshawa rule, it began to decline under the administration of Bajirao II in the latter part of the eighteenth century. Despite these fluctuations, SV Divekar notes that during the Peshwa administration, agricultural productivity per hectare was comparable to that of England, and peasants of western India, especially those living in Gujarat and Deccan,

---

5 H. Fukozawa, *The Medival Deccan–Peasants,* social systems and states, 16th to 18th centuries, Delhi, 1991, pp. 134–135.

6 Cited in... H. Fukozawa, *The Medival Deccan–Peasants*, Social Systems and States, 16th to 18th centuries, Delhi, 1991, pp. 134–135.

7 Andrew Wink, *Land Sovereignty in India*, CUP, 1986, pp. 342–343.

were no worse off than their European counterparts.[8] However, despite constant attempts by rulers to expand agriculture, a persistent shortage of manpower, coupled with the lack of technology and limited modes of transportation, rendered them unable to bring large tracts of forestland under cultivation.

Gardens and Forests

The lack of centralised control over forested areas was perceived by many as being synonymous with the lack of conservation and management principles in pre-British Maharashtra. Though there is no evidence of an independent Forest Department, woodland management practices were not unknown to the people. Rather than a systematic application of such principles, this knowledge filtered down to farmers through custom, tradition and religion.

The Konkan coastline presented a pleasing picture of gardens, orchards, fields and forests due to abundant rainfall. The soil near the sea contained marine deposits and sea sand was preferred for garden crops. Gardens and orchards of the Konkan produced such varied produce as mangoes, coconuts, betel nuts, betel vine, sweet potatoes, areca nut, palm, pineapple, jackfruit and an assortment of vegetables. Gardening was the main source of income for many families in Konkan. The Marathi saint Ramdas, a contemporary of Shivaji, gave a detailed account of gardening in his chapter called *Bag Prakaran*.[9]

The Western Ghats contained large tracts of forests with a variety of trees. These forests were located on the slopes and tops of surrounding hills. The most important product of this region was teakwood, and the quality of Kolaba teak was second only to the teak of Malabar. When the British came to Bankot in 1756, most of the waterways were covered with fine teak. The ribs and framework of most Indian Navy vessels came from Bankot and neighbouring areas. Ratnagiri timber was sold in Zanzibar through Arab traders in the early

---

8 Divekar, *Finance in Maratha Administrations*, Oriental Publishers, Delhi, 1975.

9 A.R. Kulkarni, *Maharashtra in the Age of Shivaji*, Deshmukh & Co., Poona, 1969, pp. 114.

medieval period. The Marathas also had large shipyards at Malvan and Vijaydurg. Teak was used in shipbuilding as it was durable and did not shrink or corrode. Shivaji started a shipbuilding industry at Kalyan-Bhivandi due to the supply of quality timber in the area. Even the East India Company was anxious to obtain this high-quality wood. Letters from several of the Company's factories reveal this need. One missive reads

> We have hard and ticklish game to play, for the Mughal king will be highly enraged against Sevajee, should he understand that we are in peace or hold any strict correspondence with him. It might probably cause him to order some disturbances to be given in our general affairs, not only in these parts but Bengal also. On the other side we are forced to keep fair Sevajee also, because from his country we are supplied with provisions, timber and firewood and likewise your inhabitants of Bombay drive a good trade into the main which would be a great prejudice to your island if it were obstructed.[10]

Another letter states that

> As to the treaty with Sevaji we are not wanting to let him know how considerable your power is and how advantageous your commerce will be to his ports, nor doth Sevaji want a moment ("need a monitor" in your O. C. 3649) to tell him the want we have of his wood with which your island is supplied and the other emoluments you will reap by having country open to you.[11]

Shipbuilding was one of the most prominent industries in this territory during the eighteenth century. Local Hindu carpenters, a Muslim class known as *vadhas*, and a section of seafaring classes called *bhandaris* built the vessels. Parsi craftsmen excelled in building

---

[10] Letter dated 21 Dec 1962, Bombay to Company, English records of shivaji 1652–1682, (ed.), N.C. Kelkar and D.V. Apte, Shiv Charitra Karyalaya, Poona, 1931, pp. 235.

[11] Ibid, Bombay to Company, 14 June 1672, pp. 218–219.

European-style vessels. Major shipyards could be found in Surat, Bombay and Daman. Bombay's status as the chief port of western India in the nineteenth century, along with the government sponsored migration of craftsman to Bombay, resulted in the decline of the shipbuilding industry in Surat. Bombay's supply of teak, as well as its abundance of skilled labourers and fine port, made it one of the prime shipbuilding centres in India. At the beginning of the nineteenth century, six firms of Parsi shipbuilders in Bombay monopolised the docks. Some Parsis went to England as early as the nineteenth century to study shipbuilding. From 1736 to 1857, 267 seafaring vessels, ranging in size from 23 to 2,298 tons, were built at Bombay. Of these, four were more than 2000 tons each and 30 more than 100 tons each. Among the more notable ships built at Bombay were the *Seal* and the *Buckinghamshire*.

Thana, Kalyan, Bhivandi Alibag,Vijaydurg and Malwan were the shipbuilding centres of the Maratha government under the Peshwas. Contemporary records suggest that Suvarnadurga was another centre where warships were built. There is also mention of Peshwas providing facilities to the ship-owners. The owner of a ship was exempt from *octroi* (a local tax) and other levies when it loaded goods for household use and brought them from other ports to the city. Such a vessel was also exempt from forced service to the government.[12]

Besides teak, other important varieties of trees found in these forests were *amba, ain, nana, hedu, kalamb, asan savi, jambul, sisu, kinjal, patualva* and *hirda mahua* trees, to name but a few. *Amba* wood was well-suited for agricultural implements. *Ain* was considered equal to teak for construction work. *Nana*, being a strong wood, was best suited for building houses. *Kalamb* was used for light construction work. *Asan* and *Jambul* were used mainly for construction work. *Sisu* and *sisam* were the finest quality of wood used for high-quality furniture. The wood which was not used for construction work or for making implements was typically used for fuel. Bamboo was widely used in

---

[12] R.D. Choksey, *Economic life in Bombay Konkan*, 1818–1939, Phoenix Publisher, Bombay, 1957, pp. 137. Also see Imperial Gazetter of Bombay Presidency, Kolaba and Janjra, Vol. 11, 1883, pp. 17–18.

building. Baskets, trays, and other articles were made from bamboo. Wax, honey, lac and gum were raised in the forests. The forest also supplied spices, and medicinal herbs such as avia and hirda. Leaves of certain trees were used for making Indian cigars (*beedies*) while others were used for preparing baskets, leaf plates, rain covers, and so forth.

Maratha rulers took great interest in developing plantations. A large amount of uncultivated land was cleared for the purpose of establishing plantations, and the government encouraged this enterprise by awarding the owners a wide range of tax exemptions. Coconut and areca nut plantations as well as mango orchards were established in the Konkan region, while fruit plantations were begun in *Desh*. Shahu Maharaj, a grandson of Shivaji, was particularly interested in collecting different species of plants, and he often obtained them from distant places.[13]

The importance of trees for political and defensive purposes was also evident. *Ajnyapatre*, which is the collection of Shivaji's administrative decrees, underscores this concern. Shivaji envisioned an ideal fort which emphasised on heavy afforestation in and around the fort. He said that such afforestation contributes to the defence and safety of the fort, and that these trees could be used as timber and fuel wood in times of emergency. Shivaji's instructions to his naval staff are even more revealing

> Teak wood useful for shipbuilding should be cut only after the permission of the local officers. In addition to this if some more wood is required it should be imported from a foreign country. Though the wood of mango and jackfruit can be used for the same purpose they should not be cut as people grow these trees with great care looking after them like their own children. Also, once these trees are cut, it takes a very long time for the tree to grow again to the same stage. However, in case a tree is nearing death, it can be cut, but only with the permission of the owner. Otherwise the blame of exploiting the subjects will fall upon

---

[13] Bhave, V.K., Peshwekalin Maharashtra (Marathi), ICHR, New Delhi, 1976, pp. 140–147.

> the king and his staff. And a rule that causes misery to the subjects will not last for long.[14]

As the founder of the Maratha kingdom, Shivaji's efforts to establish his legitimacy are seen through his various policies and directives to his officials. These policies underwent changes as the Maratha rule strengthened so that by the Peshwa period one finds that government imposed restrictions such as prohibition from harvesting teakwood from the village common lands. This was likely due to the growing importance of teak wood for shipbuilding and the increasing demand for teak by Europeans. We also find orders issued to preserve forests at the foothills of forts.

Shivaji's letters indicate that wood used for construction purposes from the village common land had to be purchased from the village officials. Duties were also imposed on purchased wood. In case of calamities, such as houses catching fire or destruction during wartime, wood was provided free for repair and reconstruction.

Only a nominal duty was imposed on jungle produce such as wax, honey and gum. There was no tax on firewood for self-consumption. A tax was levied on artisans who required firewood in their occupation; this tariff was called *ranva*. There was also a tax charged – as *chawal* instead of *ranva* – for the fuel wood bought from the Government *Kurans* (pastures). This *chawal* was charged per bullock. During this time, the government introduced a new system of control over jungles and pasturelands. As the breeding of good horses was crucial to Maratha success in the eighteenth century, the *silardars* (cavalry) were allotted a certain proportion of jungle to pasture their cattle. *Silardars* were contract soldiers who bred colts in peacetime. It is unclear if they had absolute rights over the grasslands and jungles they were given.

Maratha rulers controlled the use of some pastures and levied tax on wood. There were instances when people were fined if they cut wood from government pastures. The Marathas had a complex system of obtaining fodder by levying a *begar* (forced labour dues). Villagers

---

14 Ramachandrapant Amatya Ajnyapatre: Shivaji Rajanchi Rajneeti (Marathi), (ed.), N. Banhatti, Vividh Prakashan, Poona, 1960, p. 38.

had to bring fixed amounts of fodder from the government pasture to state-owned stables. An official named Balaji Krishna was appointed in 1773 to administer pastures belonging to government as well as private individuals within a 7–12 mile radius of Poona. He was to deliver to the government in Poona 1600 *khandies* (seers, where 1 seer is nearly 1 kg) of fuel and 150 seers of coal.

Pastures were sometimes taken from private individuals. A quantity of products sufficient to meet individual needs was kept by the worker, but the remainder went to the government. The area under the purview of Balaji Krishna consisted of 8 pastures and 175 foresters – 12,450 rupees per annum were paid by the government for the purpose of upkeep. A duty was levied on all the traders carrying wood on bullocks at the rate of 4 *annas* (16 *annas* = one rupee) on every rupee worth of wood. Accounts of these incomes and expenditures were delivered to the government. Such activities as grass cutting used forced labour, as they had in the past.[15]

The village wasteland included pastures. Professional grazers such as *dhangars* and *gavalis* (but not villagers) were charged a grazing fee in village pastures. The fee was levied at the rate of 6 rupees per 100 sheep.

When the interests of forests became contrary to those of agriculture, the rulers gave priority to cultivation. In a letter that instructs an official of Nago Thana Ashtami in *taluka avchitgad* to bring forest land under cultivation, it is said that

> ...orders were formally issued to prevent the cutting of timber. This prevented the cultivation of wasteland on which assessment had already been imposed. Only valuable timber forests should be preserved. Timber elsewhere should be left open for cutting. Orders were issued accordingly to the officers.[16]

---

15 Several letters are found in Chitnisi Daftaratil Kagad Patranchi Varnatmak suchi, Peshwa Daftar, Abhilekhmala Varnatmake Suchi, Vol.1–3, Government of Maharashtra, Bombay, 1991.

16 Rao Bahadur Ganesh Wad, selection from Satara Raja and the Peshwe Diaries, Government of Bombay, 1911, Vol. 6, Letter No. 716, p. 223.

However, this kind of control over the use of forest resources was not imposed uniformly throughout the region. Only in places where the government could have strict oversight – such as areas near forts or those close to the capital city – were the rules carefully observed.

Culture, Religion and Nature

Local governments of the period enjoyed a considerable degree of autonomy. Whomsoever the ruler, the villagers led relatively normal lives. There was little governmental interference in the day-to-day life of the people. Welfare measures such as donations to religious centres, construction of lakes and small dams, and the waiver of taxes in times of calamity were often undertaken by the rulers.

Important decisions relating to the affairs of the village were made by the village *panchayat* (council) itself. This council consisted of all caste representatives and was headed by the *patil* or *khot* (who belonged to the upper castes such as Maratha or Brahmin, and his assistant *kulkarni*, who was invariably a Brahmin). Problems concerning the castes were solved by the caste councils. The political, social and religious matters were dealt with at the local level.

The caste system was based on the division of society into hierarchical groups. These caste distinctions were based primarily on a person's function, and led people to develop specialized skills. This system insulated people from competition posed by others from outside their community. This in turn led to a sustainable use of natural resources as each group of workers had a right to use particular natural resources. An example which illustrates this is the rule quoted earlier that forbade butchers from stocking sheep. Instead, they had to procure them from the shepherds when needed.

Local chronicles such as the *Chitnis Daftar* cite several disputes among different communities that arose when the function of one community was performed by another (especially when it was a ceremony or a religious function). In such cases, the verdict often went in favour of the traditional practitioners of that profession. Conflicts between *upri balutedars* and *watandars* or *mirasdar balutedars* were a matter of common occurrence.

Although the exact nature of the relationship between the Hindu religion and the modern environmental movement is debated by scholars, it cannot be denied that some religious traditions facilitated the conservation of natural resources. The belief that certain trees and plants are sacred or the yearly *puja* (worship) of the Banyan tree, are but a few examples of the practices, which bring humans into alignment with their environment. The image of gods such as Munjaba and Bhairoba under a peepul tree is common in western Maharashtra, particularly in the plains of Nira valley in the Poona district.

Concurrently, the use of *belpatri, durva* (a type of grass), and various species of flowers required in the *pujas* for veneration of a divinity's idol, made it essential for people to grow such trees. The choice of flowers and plants also depended on their seasonal availability, and specifications regarding their use have been passed on from generation to generation. Each of the *pujas,* such as the *Mangala Gouri Puja, Ganesh Puja* and *Shivlinga Puja,* has its own specifications regarding the use of different leaves and flowers. In this context, a description of the routine *puja* performed by a pious Hindu is very revealing.[17] It is interesting to note that most of these plants also have some medicinal value, something accepted by people even today. For example, descriptions of households in Marathi literature indicate that planting a neem tree in the compound of a house was a common practice.

In a study entitled *A Cultural History of Maharashtra and Goa,* the author cites names of several villages and towns, which are derived from plants and trees. The banyan tree seems to be the most popular, as a large number of places take their names from it. Twenty locations are named after the fig tree alone; 11 places derive their name from the tamarind tree. This indicates the importance of sacred trees and the place of nature in society's consciousness.

Vrikshayurveda is an ancient Sanskrit text on the science of plant life, which discusses such topics as the procreation, preservation and treatment of seeds before planting; the preparation of pits for saplings;

---

[17] One of such Pujas has been discussed in great detail in the Imperial Botany Gazettear, 1887, see Appendix I.

the selection of soil; methods of watering; and the layout of a garden. Along with the enumeration of various types of trees and plants, it notes the importance of growing trees. The introduction states

> "A person is honoured in Vaikuntha (Heaven) for as many thousand years as the days he resides in a house where tulasi is grown;
> By planting four palas trees a person doubtlessly obtains the fruit of Rajasuya sacrifice;
> By planting one asvattha, one picumanda, one nyagrodha, ten tamarind trees, the group of three, viz., kapitha, bilva, and amalaka and five mango trees, one never visits hell."[18]

The Bhakti movement of the medieval period also emphasised the respect for nature shown by inhabitants of this region. Tukaram and other leading figures of this movement held that all living beings, be they human, plant or animal, were equal before God. Bhakti teachings instilled in the people a love for nature. The literature of the period is full of similes and metaphors borrowed from nature.[19]

While the planting of trees was performed largely by individuals, conservation efforts were also undertaken at the community level. Sacred groves are one such example of community conservation. Much has been written about sacred groves, which were located mainly in remote hilly areas of Western Ghats. Local people believed that spirits in the forests would take offence if any of their trees were cut, flowers or fruits taken, or animals hunted. Thus, the plant and the animal life in these groves remained unmolested.

The sacred groves are clearly discernible, as the well-preserved vegetation in them is quite distinct from the surrounding areas. They may range in size from a small group of trees to a large forest.

---

[18] The manuscript of Vrikshaurveda by Surpala was procured from Bodleian Library, Oxford, UK, Dr. Y.L. Nene. For further details, see Surpala's Vrikshaurveda, translated by Nalini Sadhale, Asian Agri-History Foundation, Secunderabad, 1996.

[19] For example, the famous Marathi saint Tukaram is one of his Abhangas says *"Vrikshavalli amha Sorjari Vanachari"*, ..., (all the inhabitants of nature like trees, creepers and the wild life are our relatives...).

Occasionally, they spread over as much as 100 hectares of land. Even the smallest grove often has old and magnificent trees. They support many species of plants, which are becoming increasingly rare elsewhere due to continued deforestation, and are often the last refuge for arboreal birds and mammals, especially monkeys, and other forest-living animals. These groves offer excellent opportunities for field research to professional botanists, naturalists and conservationists.

Many sacred groves possess perennial water sources (springs, ponds and tanks) which meet the water requirements of many villages. The age old system of every village having a temple, a tank and a sacred grove illuminates the effective water harvesting and management practices prevalent during those days. Usually, wells near the coastal areas supply only slightly saline water. However, the mere presence of groves in coastal areas ensures a supply of fresh water.

The water retention capacity of sacred groves was also well realised. The thick canopy of the sacred groves has different layers, which reduces erosion from rainfall. The water filtering through the humus seeps underground and is stored as groundwater. In short, sacred groves help to maintain a steady watertable and prevent desertification. Some groves located in laterite soils do not generally support the growth of plants; however, humus formed by gradual decay of leaves and other organic matter forms a blanket over the soil. The rich layer not only sustains the grove, but also increases the fertility of nearby agricultural lands. In addition, these groves supply raw materials to villagers, thus permitting them a livelihood.

The existence of undisturbed pockets of vegetation is mostly due to certain taboos derived from mystic folklore. Most of the tribal populations lack the ability to read and write, yet have scrupulously preserved their customs, rituals, and ceremonies through oral tradition, which often suggests appropriate behaviour in everyday matters. The folklore bestows rewards and blessings for good behaviour and prescribes severe punishments for violators. For instance, it is believed that any felling of trees in these groves would result in the death of those who cut them. Hence, traditional beliefs and customs contributed to the conservation of nature.

Compartmentalised thinking and approaches to the study of any discipline has been a prominent feature of modern society. This has led to a compartmentalisation/fragmentation of culture and with it our worldview. Consequently, the connection between socio-cultural attitudes towards nature and the management of our resources, is scarcely acknowledged and seldom explored. Resource exploitation has instead followed a pure profit maximisation approach. Despite this fact, until recently the religious and cultural dynamic played a major role in shaping Indian beliefs and attitudes towards nature.

This discussion examined how people of the Maratha region in India associated themselves with nature and managed their natural resources. On the part of the ruler, one can clearly discern a pattern of increased utilisation of these resources from the beginning to the end of Maratha rule. Shivaji, the founder of the Maratha Empire, struggled to legitimise his young kingdom, and was acutely aware of the needs of his subjects vis-à-vis the needs of his army or navy. This meant that he was less controlling in the sphere of environmental issues, whereas the later expansion of the Maratha Empire saw increased control over natural resources.

Like other self-sustaining economies, the rulers and people of the Maratha territory influenced and altered their environment in the way that was best suited to them. However, the utilisation of resources was limited by not only the rudimentary technologies and modest population, but by the attitudes and consciousness that shaped the socio-religious and cultural character unique to the region. The socio-economic structure of that society, with such unique features as the *balutedari* system, enabled the sustainable use of natural resources. Religion, of course, was a major guiding force in the society's relationship with nature, and the belief that all of nature was interrelated was the guiding principle of its spiritual ecology. As Partha Chatterjee said, "The very nature of peasant consciousness, the apparently consistent unification of an entire set of beliefs about nature and about men in the collective and active mind of a peasantry, is religious. Religion to such a community provides ontology, an epistemology as well as a practical code of ethics, including political ethics."[20]

[20] Chatterji, Partha, *Agrarian Relations and Communism in Bengal*, 1926–1935, (ed.), Ranjit, Guha, Subaltern studies, OUP, Delhi, 1982, pp. 31.

3

Early British Management of Forests

The areas along the coastal belt and Western Ghats in Bombay Presidency were once thickly forested with abundant greenery. So much so that Captain Dickenson of the East India Company stated "The whole country (coastal region) was lying waste and unpopulated. That upto about 1850, wasteland was everywhere so abundant as to create a feeling of despair as to the future of the district, that the increase of cultivation was so much desired that the poorest people were allowed to cut down as many trees as they liked merely for the purpose of clearing the land and that wood itself was abundant, that everyone cut where and as they liked."[1]

Likewise, available records indicate that the teak forests allured the British officers of the East India Company, when they arrived at Bankot, in Ratnagiri district in 1798.[2] The teakwood found in this region was of a harder grain. Being equal in size with the Malabar timber, it was highly valued of the two because of its reputed durability.[3] The curved teak logs, known as Bankot knees, were largely exported to

---

1 *Bombay Forest Commission Report*, Vol. I, Bombay, 1887, p. 21.

2 R. D. Choksey, *Economic Life in the Bombay Konkan 1818–1939*, Bombay, 1957, p. 137. The teak forests of Bankot, a part of the coastal Konkan in Ratnagiri district, were dense. Bankot is a place in Ratnagiri, a part of Konkan in the coastal region of Maharashtra.

3 George W. Sterile, *New Source of Revenue*, London, 1878, p. 84.

Mumbai. Stout ribs and frameworks of the old Indian Navy also came from Bankot.[4]

There is hardly any evidence of conflict between agriculture and forests during the pre-British times, as land was more than adequate for a sparse population. In fact, even after 20 years of British East India Company rule, descriptions of rich teak forests are found in Bombay Presidency. For instance, following are the minutes written by Colonel Jervis, the Chief Engineer at Mumbai in 1843, in his capacity as the member of the Military Board

> "...Southward from Kolwan, beyond Aptie, there is a good deal of teak, formerly planted and conserved by the Angria family. This extended eastwards to near Jambulpara and formed a large source of supply for the Poona market. In 1843 the trees were not of great size. In the taluks of Nagotna, Bohee Goregaum and Nizampure belonging to the British Government there is a good deal of teak. In many villages, such as Putwoos, it has been long preserved with the exception of cutting allowed by the Collectors and their assistants and by the Kamavisdars. Formerly, a man petitioned for leave to cut down trees for a particular purpose, chose the best and through connivance of subordinates helped himself most liberally. This in a great measure tended to the destruction of the forest. Further west in these taluks, the teakwood seems to have been cut down annually for burning in fields. Since the stoppage of teak cutting three or four years before 1843, under the orders of Government, the hills have become covered with thriving young teak trees, which in the course of 10 years or 15 years may be expected to afford a large source of supply for smaller wood, applicable to the houses of the

4 The Marathas had ship building yards at Melvin and Vijaydurg and showed a prudent regard for forest preservation. However, after the transfer of Ratnagiri from the Peshwa in 1818, cultivation increased considerably and large parts of the district were laid bare. *Imperial Gazetteer*, Bombay Presidency, Mumbai, 1908, p. 152.

> people. On the Bankot there are several plantations of teak. In 10 or 12 villages of Ryeghur taluk there are rather extensive plantations just beginning to recover from the cutting formerly allowed under the contract system, which permitted a man on paying a certain sum to have the privilege of cutting within certain boundaries for so many years. Further southwards of Bombay Presidency there is a description of two extensive plantations, which were grown by the providence of Angria family. One was near Suvarnadurg taluk and the other near northeast of Malwan...[5]

As seen in the previous chapter, during the eighteenth century, the Maratha rulers were known to have made efforts to conserve teak forests. No felling was allowed in government *kurans*.[6] However, with the exception of teak, there was no prohibition on cutting other species in the forests except in government *kurans*.

Even in Poona district, where the capital of Peshwas was situated, description of abundant forests is available. For a long time, according to the *Imperial Gazetteer of Poona*, even after the beginning of British rule, the comparatively small population and limited area under

---

5 Angria family was a Maratha naval power on the Konkan coast during the eighteenth century. The family is credited with growing several teak plantations along the coast within their territory; W. R. Baillie, *Summary of Papers relating to the Madras and Bombay Forests*, Bengal Secretariat, F. Carbery, Bengal Military Orphan Press, 1851, pp. 250–252.

6 *Selections from the Satara Raja's and Peshwa's Dairies*, Vol. 1–9 (Marathi), Bombay/Poona: Poona Deccan Vernacular Translation Society, 1906–1911. There are other records like *Peshwe Sannika*, from Peshwa Dafter, Poona, which mentions prohibition on felling of timber from government *kurans*; Also the *Bombay Forest Commission Report* reports that new taxes had been imposed by the Maratha rulers in certain areas towards the end of the eighteenth century. These taxes were like *Vancharai* or grazing tax, duty on export of hay, straw and ghee. *Vanamakta* was a general duty on cutting wood from the hills which were leased out to farmers. The letting out of "hay grounds" was also one of the new items of revenue introduced by the Marathas, *Bombay Forest Commission Report*, 1887, Vol. 1.

cultivation made any special measures for preserving forests unnecessary. How to clear for tillage the large area under trees and brushwood, was one of the most pressing administrative questions of the time. This was mainly the objective when (1836–37) revenue survey was introduced in west of Poona district. Almost all hillsides were divided into plots and offered at nominal rents. It was about 20 years later when the population increased leaps and bounds and the railway was opened through the Bor Pass, the demand for wood increased. This increase in demand for wood combined with the ease with which it could be sent to the markets began to strip the country of its forests. To check this, certain lands were set apart by the government as forest reserves. In 1849, demarcation of forests was started by Dr Gibson, the father of Bombay forestry. He chose forests of tree-covered land which the people still call *Daktari Ran* or the Doctor's forests.[7] This description shows that at the time when British took over the territory from Marathas the state of affairs, as far as forests were concerned, was rather satisfactory as against the constant criticism made by the latter British officers. In fact it was the British administration at the beginning of their rule, which caused the degradation of forests of this region.

From the beginning, the British policy was to extend agriculture by bringing more lands under cultivation, the objective being increasing revenue collection. The direct and indirect value of forests were underestimated. Forests were believed to be an obstruction to agriculture in many places, and therefore a limiting factor to the prosperity of the country. In fact, according to Zimmerman, "British believed jungles to be lands that had lapsed into a state of nature because of inadequate care of man. The jungle was seen as a result of the abandonment of cultivation and was a place of wild, rank vegetation".[8] The extension of the cultivated acreage was an index by which the British evaluated the success or failure of their policies. The

---

7 *Imperial Gazetteer*, Poona district, Vol. 15, 1885, Part II, p. 31.

8 Zimmerman, *The Jungle and the Aroma of Meats: Ecological Themes in Hindu Medicine*, London, 1989, p. 14, 44, cited in Mahesh Rangarajan, *Fencing the Forests: Conservation and Ecological Changes in India's Central Provinces 1860–1914*, OUP, Delhi, 1996.

forest was a landscape to be conquered and tamed.[9] As such large forest areas got transferred to landowners or cultivators, though without the conferment of any legal rights. In other cases where the forests were not entirely alienated, the main rights of users, which constitute the value of possession of forests were abandoned in favour of the cultivator. This, according to people like E.P. Stebbing, was a transitionary phase. But it lasted for many years causing an enormous destruction to forests.

The need to stop the mindless destruction of forests began to be felt only when the demand for steady supply of timber for shipbuilding arose.[10] Formerly the Royal Navy had depended on oak for its shipbuilding requirements. However the oak plantations in England were almost exhausted by the end of the nineteenth century. Teak from India was the next best alternative.[11] Hence the solution was sought for in improving methods of exploitation – both by government agencies and contractors. However, even when protection was accorded, for many years it was given only to certain species of trees and not to the forests as a whole.

The first concrete step taken by the government towards forest protection was an order issued by Bengal-Bombay Joint Commission, appointed in 1800, to inquire into the internal circumstances of Malabar and to make regulations prohibiting the felling of teak below 21 inches girth.[12]

In 1805, a dispatch was received from the Court of Directors of the East India Company inquiring about the extent to which the might of the King's Navy, in view of growing deficiency of oak in England, depended on a permanent supply of teak timber from Malabar.

---

9 *Ibid*, p. 18.

10 B. Ribbentrop, *Forestry in India*, Delhi, 1900, reprint 1989, p. 67.

11 R. G. Albion, *Forests and Sea Power*, Harward University Press, Cambridge, Massachusetts, 1926, pp. 367–368.

12 The earliest record of attempts to solve the timber problem is the formation of a timber syndicate in Malabar in 1796 with Machochie of Medical Service as the leading spirit. See B. Ribbentrop, *Forestry in India*, Delhi, 1900, reprint 1989, p. 67.

Consequently, a Forest Committee was appointed to enquire both into the capacity of forests themselves and the status of proprietary rights to them. To maintain its supremacy on sea against the French, it was necessary for England to get over this problem of shortage of timber for shipbuilding. The Court of Directors was aware that Indian timber had long been prized and that in fact, Arabs had imported teak from Bombay for building their fleet. It was Bontius who was the first to describe Indian timber under the name *Qurecus indicus.*[13]

The reports submitted by the Forest Committee appointed by the Court of Directors indicated that the more accessible teak forests of peninsular India had almost been exhausted. Tapping the distant forests would require construction of costly roads. Nevertheless, the Committee pointed out that if protection was afforded, these forests could be gradually regenerated. This highlighted the necessity to protect forests along with the improved methods of utilisation. The immediate result of the Committee's report was that in 1806, an officer was appointed as a Conservator and in 1807, a general proclamation was issued, that the royalty rights in teak trees proclaimed by the former government in the south of the Indian continent was vested in the Company, and all further unauthorised fellings of the teak were prohibited.[14]

The instructions of the Court of Directors stated that their objective was to receive a regular supply of timber for public purposes, from lands that were not appropriated. It was not designed to invade private property or to prevent anyone from disposing of at his pleasure the timber on his own estate. In the new set-up, however, the Conservator, assumed much larger powers than were entrusted to him

---

13 By the time the committee was appointed, it was a well-known fact that the planting of oak owing to supiness of successive governments had fallen into abeyance for nearly a century, and England was faced with shortage in timber supplies. See E. P. Stebbing, *Forests of India*, Vol. I, London, 1926, p. 63; Also see W. R. Baillie, *Summary Papers*, Relating to Madras and Bombay Forests, Bengal Secretariat, 1851, p. 178.

14 W. R. Baillie, *Summary of Papers*, Relating to Madras and Bombay Forests, Bengal Secretariat, 1851, pp. 178.

earlier. On 10 November 1806, Captain Watson of the Police Department was appointed as the first Conservator of Forests in India.[15]

However, the proclamation of 1807, which vested authority in the hands of the Conservator, did not define the term 'sovereignty'. Nor were the forests specified over which such sovereignty extended. But Captain Watson, acting upon his views on these points, succeeded within a short time in establishing a monopoly over all the timber in the two provinces of Malabar and Travancore. He cut down and appropriated for the use of the Company not only trees in private forests but even those grown on cultivated lands, that paid revenue to the government. Even the proprietor himself, unless expressly permitted by the Conservator, was prevented from cutting a piece of wood on his property or removing the young seedling plants, though they might be injuring his land. The immediate successors of Captain Watson followed the same policy.

Consequently, trade in timber was almost annihilated. Even if a merchant could obtain the Conservator's permission to purchase private timber not required by the government, he no longer had the same means of disposing it, its expropriation having been prohibited soon after the establishment of the new system. Also, the peasantry were deprived of the privilege of cutting wood for fuel and other ordinary purposes – a privilege they had enjoyed from times immemorial. The way these measures were implemented made the system intolerable, even causing a lot of discontent among the proprietors and merchants. This feeling rose to such a pitch that on the recommendations of the Governor of Madras, Sir Thomas Munroe, and with the consent of the Supreme Government, the conservatorship was abolished in 1823.[16]

Sir Thomas Munro was very much against any sort of interference in the people's rights over forests. He said "They (merchants and

---

15 This officer was supposed to have a good knowledge of the forests as also the language and habits of the people. See E. P. Stebbing, *Forests of India*, Vol. 1, London, 1926, p. 64.

16 W. R. Baillie, *Summary Papers*, relating to Madras and Bombay Forests, Bengal Secretariat, 1851, pp. 179.

agriculturists) were 'too good' a traders not to cultivate teak or whatever wood that is likely to yield a profit. They are so fond of planting ... to encourage them no regulation is wanted, but a free market. Restore the liberty of trade in private wood. Let the public be guarded by its ancient protector, not a stranger. But as the Collectors and Magistrates of the country, we shall get all the wood the country can yield, more certainly, than by any restrictive measures. Private timber will be increased by good prices and trade and agriculture will be free from vexations."[17]

According to E.P. Stebbing, Inspector General of Forests "This pious hope showed a complete ignorance of the point of view of the private proprietors of forests, timber merchants and their methods of working. It sounded the death knell of hundreds of thousands of acres of fine teak forests."[18]

After the abolition of Conservator's office and the earlier restrictions, the forest holders, or the persons who obtained from them or from government permission to fell timber, bothered only about their interests. Probably as a reaction to the prohibition, they began to completely clear the land, indiscriminately cutting all trees without leaving any young plants for regeneration. As a result, some of the more accessible forests were destroyed or were materially injured.

Soon after the introduction of British rule in Bombay Presidency, the Konkan forests were treated solely as depots for supplying knees

---

17 E. P. Stebbing, *Forests of India*, Vol. 1, London, 1926, p. 84; Sir Thomas Munroe skillfully exploited the favour that Mount Elphinstone enjoyed among directors by extolling the policies that were being pursued in Bombay, which were modeled on his policies in Madras, and being carried out by Chaplin with whom Munroe kept in close touch. He was confident that influential friends would pass along his approbation of Elphinstone's procedures to others. He also suggested that the modification of the Madras system in Bombay provided valuable pattern for the entire British India. See Burton Stein, *Thomas Munroe: The Origins of the Colonial State and His Vision of Empire*, OUP, Delhi, 1989, pp. 326–327; Also see Sushma Varma, *Mount Stuart Elphinstone in Maharashtra : A Study of the Territories conquered from the Peshwas*, Bagchi and Company, Calcutta, New Delhi, 1981.

18 Ibid, p. 85.

and other timber for shipbuilding. However, very few efforts were made to improve the growth of teak in special areas by maintaining a small establishment or by encouraging the private proprietors to do the same. For example, a letter from Anderson Dady, in 1810, shows that he promised to follow government's recommendations of preservation and improvement of valuable timber that were there within his boundaries.[19]

However, it was only after 1818 (the Battle of Plassey), that the British consolidated their position in the Maratha territory and comprehensive measures to improve teak plantations were taken up. In 1823, Dunlop, the Collector of the Southern Konkan comprising Kolaba, Sanski, Rajapuri and Raigad, recommended to the government that the right to teak and other valuable trees be given to the *rayots*.[20] For, as Dunlop has pointed out, only two teak forests in Suvarnadurg and Malwan taluks, were carefully preserved by the previous government. They contained almost as much teak as the space could permit. However, in other parts under the jurisdiction of his Collectorate, there was scarcely any such thing as teak tree of any size. He gave two reasons for this. One was that the native governments had appropriated all teak trees, which were worth taking. The other reason he gave was interesting. To quote

> "In the first place Konkani agriculturists entertain an opinion that without burning grass, boughs of trees and in short all vegetable substances they can get on their fields, no tolerable crop is ever produced. This practice is universal. Portions of hill producing jungles are therefore assigned to each field or division of fields, for the purpose of supplying and particular plants are known to the husbandmen to be much superior in their effects to others; teak is not only the worst of the whole for this purpose, but wherever it becomes once established destroys all others. The teak tree sheds its leaves at an earlier season

---

[19] Early History, *Bombay Forest Commission Report*, Vol. I, Government Press, 1887, Bombay, p. 18.

[20] *Ibid*, Vol. I, pp. 408–411.

than others and leaves soon become so wasted by the weather as to retain but little substance and neither the leaves nor the wood are found to be of almost any use for burning on the fields from some property of the tree, also grass and other plants do not grow under it, so that altogether it is extremely unpopular with the cultivators and would never be suffered to grow except for the qualities of its timber. For these reasons, it is vain to hope that we should ever be able (even if it were desirable, which is doubtful) to offer sufficient encouragement to induce the natives to grow teak except where there is more hilly jungle than they require for the use of their fields and in such circumstances there seems little doubt that the measure proposed by Sir Thomas Munroe of declaring all teak to be the property of any individual who might grow it, would in the process of time lead to the wood being preserved which is all that is necessary, for teak springs up spontaneously almost all over the Konkan..."[21]

Dunlop in his communication to the Government forcefully brings out the fact that the British concept of forests was dominated by a policy of emphasis on commercially valuable species of trees like teak. From the Peshwa period onwards teak seems to have assumed greater importance for the rulers, and hence it was looked upon as a 'Royal Tree.' One can infer this from Dunlop's letter, though the common people's apathy towards this particular species is understandable. However, accusing common people of enormous destruction and deforestation seems unreasonable as the letter further states

"...the present forests are quite recognised as the government property and are not liable to the principle objection urged against the government claims in Malabar of being undefined. I therefore think that the established forests should be retained by government. And with the hope of encouraging the growth of teak a proclamation

21 *Ibid*, p. 21.

> should be issued surrendering all the claims asserted by former governments to teak or other valuable woods and declaring that wherever these may be found beyond the limits of the three forests of Band, Tadil and Vinhere in the Suvarnadurg taluk and of Mahan near Malwan they shall henceforth be the full and exclusive property of the persons on whose grounds they may grow, who shall be free to cut or dispose of such trees in any way they choose."[22]

In a reply to this letter in 1823, the Government of Mumbai authorised Dunlop to issue a proclamation surrendering all claims to teak or other valuable wood beyond the limits of the forests of Band, Tadil and Vinhere. However, this decision of surrendering claims to teak or other valuable woods was regretted by the administrators for a long time afterward. For instance, A. Elphinstone, the Collector of Ratnagiri, in his letter to John Vibhrat, Esq., Revenue Commissioner, Mumbai, pointed out that private individuals abused this proclamation. Also, the permission granted to the Engineering Department to cut wood for construction of public buildings etc. had thoroughly exhausted the forests in this district.

During his time, Kanhoji Angre, a Maratha chief admiral, ordered the villagers to sow the seeds to raise timber for his cruisers. As a result, there were fine teak plantations covering an area of around 2,000 bighas which led to the conquest of this district by the British. Subsequently with the Kolhapur Raja's orders, teak plantations were grown in Malwan *taluka*, around 50 years before the British conquest. However, timber from these and Survarnadurg plantations were sold by auction by the British. It was reported that a purchaser had made 1700 rupees net profit in a transaction relating to Malwan *taluka* plantations alone. Knowing the situation, the Collector of Ratnagiri stated that these forests were so badly exhausted that for another 50 years they would not produce timber fit for the naval dockyard.[23]

---

22 *Ibid*, pp. 21–23.

23 George W. Strettell, *A New Source of Revenue for India*, London, 1878, pp. 80–83.

Dr. Gibson observed that due to this proclamation to surrender government claims to teak in Ratnagiri district, individuals in the neighbouring districts cut timber as and when they pleased and if it was teak it was always represented as teak from Ratnagiri district.[24] As time passed by, this proclamation became a subject of contention since it was necessary for an individual to prove that he was the proprietor of his land in order to be conferred any rights on the timber trees.[25]

According to the 1831 report of the Collector of Ratnagiri district, there were four preserved forests in the Konkan. Though the prohibition against cutting timber trees beyond the forest reserves was removed in southern Konkan, it was retained in the Selsette region and the government had the right to all teak trees except in villages that were made over to the proprietors.

The government earned revenue from public forests in other ways also. For example, tax was imposed till 1838 on timber and firewood removed from the jungles. It was collected by the Land Customs Establishment and was known by different names as *kolai, tijai, ramkhanda, wunee, sherista,* or *kabarmikta*. In some cases, this tax was levied in addition to the ordinary custom duties and it was meant for the privilege of cutting wood in the forests. In other cases, it was merged into the customs by deliberately fixing them at high rates. This was to include the seignorage for the privilege of felling government wood as well as the duty on export. The receipts from these sources were considerable as the rate of tax varied initially from 33 per cent to 37 per cent ad valorem.[26] In the initial years, the tax was farmed.

---

24 *Forest Report* of Bombay Presidency, 1849–50, Bombay, 1861, p. 9.

25 R. D. Choksey, *Economic Life in Bombay Konkan 1818–1939*, Bombay, 1957, pp. 138–140; Two cases of Pingonde and Surve were quoted where these *khots* had to show that they were the owners of the land for getting the rights to timber trees. For details see Early History, *Bombay Forest Commission Report*, Vol. 1, 1887, Chapter 1, p. 33.

26 George W. Strettell, *A New Source of Revenue for India*, London, 1878, pp. 76–81; Also see *Statistical Report of the Kolaba Agency*, Bombay, 1851.

However, the farmers did not charge the prescribed high rates owing to a fear of ruining their existing trade.[27]

Along with the abolition of transit and inland custom duties, the wood tax too was abolished in 1838, wherever it had been placed on the transit rather than the cutting of wood. This left a confused and inconsistent system until a uniform wood tax was decreed in 1851.[28]

In 1837, the government's attention was drawn to the indiscriminate exploitation of forests. The then Acting Secretary, Townsend, observed that "It is, however, probable that the great decrease of the forests in the North Konkan which has taken place since the English obtained possession of the country, and which is yet progressing will be in course of time the most effectual check to the trade. The carved timber and knees used in ships and large boats were obtained from North Konkan and realised a high price formerly; the Governor in Council would wish to know, whether such timber is still procurable in the Konkan; if not he conceives, it would be well to refer to the Collector of Thana to consider and propose measures which may restore and perpetuate the supply of this valuable and useful forest timber, particularly in those situations which command the readiest conveyance to the coast."[29]

The first step adopted by the Bombay Government was to issue orders in 1839, prohibiting the felling of teak without permission. In the following year Gibson, the Superintendent of the Botanical Gardens, was directed to inspect and report on the northern forests of the presidency. His report, in fact confirmed the destruction of teak forests that was taking place.[30]

---

27 Farming of tax meant selling the rights of taxation to the highest bidder by auction. See George W. Strettell, *Ibid*, pp. 76–81.

28 Richard Tucker, *Forest Management and Imperial Politics in Thana district*, Bombay, 1823–1887, *Indian Economic and Social History Review*, Vol. 16, No. 3, 1919, p. 280.

29 Early History, *Bombay Forest Commission Report*, Vol. 1, 1887, p. 39.

30 E. P. Stebbing, *Forest of India*, Vol. 1, London, 1926, p. 78.

Meanwhile, on the subject of royalty rights, Underwood, a Principal Collector of Malabar, sought to justify the royalty rights of the government over forests based on the precedence set by Tipu Sultan. He argued that Tipu Sultan had annihilated private rights in the forests and created his monopoly over them. After occupying the country of Tipu Sultan, the Company's government acquired the monopoly by succession. Underwood maintained that on the same principle the destruction of government forests by private individuals should not be allowed. He expressed the opinion that "he could not discover any record that any of the Conservators had taken any steps to perpetuate the forests or even planted a single tree." Hence, he advocated a revival of the conservatorship with abridged powers, while the Collector would be an Ex-Officio Conservator.[31]

Nevertheless, in 1840, the Court of Directors reviewed the forestry question in its entirety. It did not approve of "royalty or forest right proposal" because, if the price of timber rose it would dissatisfy the proprietors. The Court of Directors felt that there was no advantage in purchasing more land than was actually required to make provision for the government's needs in forestry materials. It ordered that the forests not required for the government's need be left to private ownership.

The Court suggested that a high rate of duty be placed on all teak felled under a certain size to deter the felling of such trees. It reiterated that the "anxious wish (of the Court of Directors) was that in the prosecution of the survey, if such a measure should appear necessary, utmost care must be taken to avoid any infringement of the rights or any unnecessary interference with the convenience of private persons."[32]

In another letter, the Court expressed its gratification for not having to act on the suggestions of Underwood and the Board of Revenue of Madras to re-assert the right of government to the

---

[31] W. R. Baillie, *Summary Papers,* relating to Madras and Bombay Forests, Bengal Secretariat, 1851, p. 204.

[32] E. P. Stebbing, *Forests of India,* Vol. 1, London, 1926, p. 86.

conservancy of all forests, whether held by private persons or made over to the government. The Court of Directors remarked that "The right of government to interfere in this manner, now that its pretensions to the ownership of the forests is tacitly renounced, seems to be more than doubtful and even if it were not disputed its exercise would be scarcely less troublesome to government than vexations to the individuals who would be exposed to it, while all the advantages that accrue from it may be obtained indirectly in a manner equally effectual and much less invidious."[33]

According to E. P. Stebbing "the Court of Directors did not recognise at that period that the interests of the Indian people were immediately bound up with the forestry question."[34] This comment of Stebbing, in contrast to the views expressed by the then Court of Directors, shows the different approaches held by the administrators during the British period.

The Government of India, on the receipt of the dispatch from the Court of Directors mentioned above, called upon the Governments of Madras and Bombay for necessary information relating to the Malabar and Canara forests.

Sending its reply in June 1840, the Government of Madras forwarded a letter by Conolly, the then Acting Principal Collector of Malabar. In his letter, Conolly expressed his opinion favouring acquisition of private forests to the extent sufficient to supply the average quantity of timber required for public services. But he thought that it would be difficult to make outright purchases of private forests considering the feelings entertained by the natives on the value of their proprietary rights and the loss of honour they would suffer if their rights were alienated. Conolly felt that the desired end could be effectively achieved by taking the forests on the usual mortgage tenure and by advancing the value of the estates. This way, he argued, the government can be secure from any intrusion. To quote "Malabar

---

[33] W. R. Baillie, *Summary Papers*, relating to Madras and Bombay Forests, Bengal Secretariat, 1851, p. 224.

[34] E. P. Stebbing, *Forests of India*, Vol. 1, London, 1926, p. 87.

mortgages are never foreclosed, but by a common tenure; the proprietor in case of redemption is bound to pay for all improvements made by the mortgagee."[35] This plan he thought would ensure perpetual possession, where the original sum advanced was equal or nearly equal to the value of property. It would also leave the proprietary rights untouched. Conolly further stated that this would "feed the harmless vanity of the proprietors, whose sole rights will be sundry seignorial ones such as the tusks of elephants that may die within the forests, the wild honey bee wax, etc., which will in no way interfere with the full occupation and use of ground and one of the greatest advantages of which will be the power of replanting the tracts that are denuded of teak."[36]

On 29 August 1840, Lord Auckland, the Governor-General, wrote a minute in which he expressed his approval of the measures that were generally taken for the prospective preservation of Malabar teak forests. But he objected to the measures of prohibition and imposition of duties, contrived to discourage felling of small trees in order to have plentiful prospective supply of long timber. He did not approve even minute interference by government. Lord Auckland thought that more good might be done if the Rajahs and greater proprietors of the woods were tied down to conditions of management, where provisions would be made to purchase their growing trees, thereby introducing a good system.[37]

During the 1840s, efforts were made to improve the forests by raising teak plantations in the principal forest regions like Malabar, Canara, Travancore, Gujarat and Konkan, giving high priority to the rights of individuals in the forests. But these attempts were not taken up on a large scale. Though it was realised that scientific advice in the management of forests was becoming an urgent necessity, no step was taken in this direction till 1845. It was only after 1845 that the Bombay

---

[35] W. R. Baillie, *Summary Papers*, relating to Madras and Bombay Forests, Bengal Secretariat, 1851, pp. 211–212.

[36] *Ibid*, p. 212.

[37] *Ibid*, pp. 213–214.

Government appointed Dr Alexander Gibson, Curator of the Bombay Botanical Gardens, as Ad-Interim Conservator of Forests. In 1847, Dr. Gibson rose to the position of Conservator of the Bombay Presidency.[38]

During his 10-year tenure, Dr. Gibson played a major role in promoting forest conservancy in India. He directed his energies towards three objectives: (1) prohibition of Kumri cultivation; (2) thinnings amongst the young teak areas in the reserves and raise teak plantations; and (3) carrying out a systematic study of the effect of denudation of forests on the climate of various localities, the water supplies in these areas together with the results perceivable in the drying up of springs and streams and silting of rivers and harbours on the coast.[39]

---

[38] The debate on conservation of forests was given a fresh impetus by the surgeons in the service of the East India Company. The surgeons, especially Alexander Gibson and Hughe Cleghorn widened the issue beyond protection of timber. From 1837 onwards they pointed to the connection between deforestation and drought. Protection of forests was now seen as essential for maintaining water supplies and safeguarding agricultural prosperity. Mahesh Rangarajan, *Fencing the Forests: Conservation and Ecological Change in India's Central Provinces*, 1860–1914, OUP, Delhi, 1996, p. 23; Also see Richard Tucker, "Forest Management and Imperial Politics in Thana district, Bombay, 1823–1887," *Indian Economic and Social History Review*, Vol. 16, No. 3, 1919, p. 280. Foresters like Gibson and Cleghorn were known as 'Desiccationists'; desiccationism was a movement which emerged in the eighteenth century, believed in the connection between the depletion of tree cover and phenomenon like droughts or soil erosion. The transcendental implications and apocalyptic nature of such concerns gave calls to halt forest erosion and urgency which went beyond the considerations of either legality or legitimacy: to fail to act to 'conserve forests was to invite disaster', For more details see, Ajay Skaria "Timber Conservancy, Desiccationism and Scientific Forestry: The Dangs 1840s–1920s" in *Nature and the Orient: The Environmental History of South and Southeast Asia*, (ed.), Richard H. Grove, Vinita Damodaran and Satpal Sangwan, OUP, New Delhi, 1998, pp. 596–635.

[39] E. P. Stebbing, *Forests of India*, Vol. 1, London, 1926, p. 220, and *Bombay Forest Commission Report*, 1887, p. 41.

Dr. Gibson also appears to have attempted demarcation of village reserves in the Northern Konkan in 1845–1846. Before him, Law Collector of Thana, during 1845–1846, assigned to each village, a timber reserve proportional to the size of the village bounds. He ordered that these reserves should be carefully preserved for the use of villagers and also the timber growing on sindad land (this is a piece of land maintained and used to collect dry leaves, branches, grass etc. to burn and make keed beds before serving. This was a common practice in Konkan region.), i.e. the ground wherein the trees or bushes grew from where the *rayots* cut the material for burning in adjacent fields. This demarcation was a very lose assignment of a small part of the wastelands of each village for the growth of timber mainly to meet the local needs.[40]

However, Dr. Gibson exercised his control and supervision over all the lands capable of forest growth or actually containing any forests. During his tours he observed that due to their proximity to Bombay, the *Watandar Khots* in north of Bankot were unusually clamorous about the extent of their forest rights. Hence he introduced the "share system." Under this system, the *Khots* or other village heads entered into an agreement with the Forest Department to conserve the teak forests around their villages in exchange of permission to lop the trees for *rab* and take one-third or one-fourth of the net proceeds during the periodical fellings.[41] Later on when the system of "revenue survey" was introduced in these villages, the Survey Department entered into similar contracts with the Khots at the time of settlement.[42]

40 *Ibid*, Also see *Forest Report* of Bombay Presidency for the years 1849–50, 1852–53, 1853–54, Education Society's Press, Bombay, 1861.

41 *Rab* was the practice of burning branches of trees, leaves, twigs, etc., on the fields in order to create beds for sowing before monsoon. See *Forest Report*, *Ibid*, 1849–50, p. 9.

42 By Government Resolution No. 7863, dated October 4, 1884, the Conservator of Forests was vested with discretionary power to renew or not, as may seem best, the Kabuliyats with the heirs of the original executors. See *Bombay Forest Commission Report*, 1887, Vol. I, p. 41.

The main sources of revenue during Dr. Gibson's time were thinnings of the teak reserves and the collections made under the fee system. It was in July 1851 that the government sanctioned the "fee system" proposed by Dr. Gibson, for levying a fee throughout the Thana district abolishing the old *kolai* tax. The fee system was introduced taking into consideration the fact that in 1836 a mistake had been made by abolishing transit duties as also many other levies made on forests. These levies were actually a seignorage and had nothing to do with the transit duties except that the same establishment levied them.[43]

But the fee system led to a prolonged clamour and strong opposition. The immediate reaction was that the merchants in Surat area organised a protest along with the Varli tribals. A description of this protest is provided by Dr. Gibson himself. To quote

> "Of such clamour we had an example in 1851, when the Gujarat merchants and others in the Sanjan taluk shipped about 300 of the Varli tribals for Bombay and marshalled them on the steps of the Council room in order to concur the government into a repeal of the jungle fee scheme which had just then come into operation...When I proceeded to Sanjan in January 1852 in order to make a final settlement I was surrounded by some thousands of persons, all of them with the same object and disposed to even personal violence."[44]

After this all forest material required for local consumption was virtually exempted from the payment of fee, and the fee was almost exclusively levied on wood removed from the forests for trade purposes. These exemptions in favour of the local population satisfied all reasonable claims, and hence there was no further opposition to the system. Dr. Gibson was also of the opinion that if possible the

---

[43] Gibson makes references to the fact that Poona Daftar record showed that the paramount government had a right in teak wood. See *Forest Report, op.cit.*, 1849–50, p. 3, 18.

[44] *Bombay Forest Commission* Report, 1887, Vol. I, p. 42.

cultivator should have some personal interest in growing forests both as a measure of justice and of future benefit to the plantations.[45]

Nevertheless, owing to the growing scarcity of timber and the extensive demand for wood for building and railway purposes, prices shot up steeply. This tempted everyone permitted to fell timber or enter the forests to make the best use of the available opportunity. As a result not only was timber removed clandestinely from the forests, the felling was done recklessly and in a wasteful manner.

In order to prevent this, in 1852, lopping of teak for *tahal* was prohibited for the first time. In 1858, a similar order was issued forbidding people from felling or even pollarding trees in their *shindad*, the area maintained for *tahal* in *rab* lands. However, in 1859 another circular was issued allowing the *rayots* to lop branches of trees other than teak and black wood and cut shrubs etc. on *rab* lands.[46]

Gibson experienced considerable opposition from the Collectors of the districts over the introduction of conservancy. The Collectors were, as a body, strongly opposed to the advent of a new department that would take over the management of the forest areas and along with them the revenue of their districts. Also, a divided forest administration in the hands of several departments led to considerable confusion with little real progress in conservancy. The revenue officials were so keen on bringing increasing amount of land under cultivation that by 1873–74 there was hardly any land left which was fit for cultivation, in contrast to the beginning of the British rule.[47]

After Gibson's retirement in 1860 and with the suggestion of Goldfinch Anderson, Collector of Dharwar, Dalzell, the Chief Conservator of Forests between 1860 and 1869 in Bombay Presidency, passed rules directing the entire forest establishment of each Collectorate to supervise the sale of timber and keep the accounts relating to the forest management of their Collectorates. The duty of

---

45 *Forest Report*, of Bombay Presidency for the years 1849–50, Education Society's Press, Bombay, p. 5.

46 Circular No. 56, *Bombay Forest Commission Report*, *op.cit.*, Vol. 2, p. 152.

47 See *The Annals of Indian Administration*, 1874–75.

the Conservators was to annually visit the forests and advice the Collectors accordingly. Also, each Collector was to be assisted by an Assistant Conservator under whose immediate supervision the forest establishments of the Collectorates would be placed.[48]

These new regulations explicitly subordinated the Conservators to the Collectors. Because of such close association with the Revenue Department, the Forest Officers more than ever came to be identified as tax Collectors and enemies, particularly in the eyes of the timber merchants. In addition, the staff of the Forest Department was grossly inadequate compared to the vast forested areas they had to manage. It lacked not only in number but also training. New recruits in the Forest Department were chosen not on the basis of effective prior training but on vague criteria of aptitude. In some instances, they were naturalists and in others sportsmen.[49] None of them had any training in the legal complexities they would face. Neither did they have the capacity for performing the delicate diplomacy of negotiating with village landholders and peasants.[50]

Hence, though the forest establishment was revised and strengthened in an attempt to have a firm control over the forest, a strict process that was also somewhat haphazard left the foresters with few friends and many critics. The stricter conservancy led to many complaints both on the part of local residents and the merchants who supplied fuel to the Bombay market.

Nevertheless, a dispatch from the Governor General in Council to the State expressed certain views, which were accepted as general

---

[48] E. P. Stebbing, *Forests of India,* John Lane, London, p. 329, and Richard Tucker, "Forest Management and Imperial Politics", *Indian Economic and Social History Review,* Vol. XVI, No. 3, pp. 281.

[49] Berthold Ribbentrop, *Forestry in British India,* first published in 1900, reprinted in 1989 with commentary by Dr Ajay Rawat, New Delhi, pp. 78.

[50] Richard Tucker, "Forest Management and Imperial Politics", *Indian Economic and Social History,* Vol. XVI. pp. 284, Richard Tucker says that even in the technical work of forest management it took a generation before the first true professional foresters were trained at the Forest Institute, which opened in 1906 at Dehradun at the Himalayan foothills.

principles in laying a foundation to the forest administration in India. The dispatch stated that

> "In the first place, under no conceivable circumstances is it possible that personal interest can be made compatible with public interests in the working of forests. The length of time required for maturing or growth of timber is so great that no individual can have a personal interest in doing more than realising the largest possible amount from any forest tract of which he may get possession. In fact timber is a produce no man can expect to get more than one crop in his lifetime. The moral and social restraints that are likely to operate to prevent such a course are most especially wanting in India, whether we deal with natives of the country or European settlers."[51]

Further, the dispatch emphatically stated "In the prevailing conditions and in this respect, personal interests are not only incompatible with public interests but they are absolutely antagonistic." It suggested that all government forests should be strictly set apart and made unalienable. Wherever private rights existed they were to be respected. "The idea", the dispatch goes on to argue that, "forest is a thing valuable in itself and in truth just as essential to the community as fields of wheat, sugar or cotton, is still not generally realised in a complete manner which is as essential before forest management can be said to stand on a proper basis ... . The subject of forestry, however, has of late been more considered and the conviction has been arrived at, that it requires the stability of a settled administration to prevent the present destruction of forests and hand them down in such quantity and conditions as to leave a due supply for future generations. A

---

51 E. P. Stebbing, *Forests of India*, John Lane, London, pp. 526–527; Similar views were expressed by a German author stating "It is only the lapse of time that can build up the forest capital; and it is because of this necessarily slow progress that the State is best fitted to manage forest estates. See W. Rosher", A Fragment from a German writer on the Reasons why the State should Manage Forests", notes and queries, *Indian Forester*, December, 1876.

permanent government only can be expected to wait long enough to reap the profit obtainable."[52]

As a first step towards carrying out the important objectives stated in the dispatch, Dr. Dietrich Brandis was summoned from Burma. The Imperial Forest Department was formed in 1864 and Dr. Brandis was appointed as the First Inspector General of Forests of Government of India on 1 April 1864. Unlike in the past, he recruited trained personnel for organising forest operations and for establishing the Forest Department in the States.

In the organisation of forest establishments before 1865, the leading principle followed was decentralisation. The management of the business in each province was left entirely to the government of that province, the Government of India reserving to itself the right of insisting upon maintenance of correct principles. The Chief Forest Officer in each province was designated Conservator of Forests, working in subordination to the local government. This officer had the control of the entire forest business in the province. At a later period decentralisation was extended downwards. In those districts where forests existed, the officers in charge of forest divisions were made subordinate to the Chief Civil Officer of that district, that is the Collector or Deputy Commissioner. At the outset it was difficult for these officers to manage the forests, for systematic forestry was a subject entirely foreign to the majority of civil officers and the only plan to ensure vigorous action in the right direction was to appoint one Chief Forest Officer in each province, who while subject to the control of the local government would take the initiative and organise the protection, improvement and working of the forests. Also, most of the civil officers of that time were utterly hostile to the interference with old established customs. They regarded the measures proposed as Utopian and an outcome of theoretical speculations.[53]

---

[52] E. P. Stebbing, *Forests of India*, John Lane, London, p. 527.

[53] D. Brandis, *Forestry in India: Original Early Development*, First Print 1897, reprinted by Samar Singh, WWF, Nataraj Publications, Dehradun, 1994, p. 127.

In January 1864, Sir John Lawrence landed in Calcutta as Governor General of India. He believed that the Chief Civil Officer should be supreme in his district and there must be no departmental interference of any sort. When Sir Lawrence was back home in 1859, as a Member of the Council of India, he regularly saw the dispatches relating to forest administration and did not approve the plan of establishing a new department that would have charge of the waste and forest lands, and that might in many ways interfere with the supreme authority of the Chief Civil Officer in his district. When Sir John Lawrence became the Governor General of India, Richard Stracey, the Secretary in the Public Works Department was in-charge of forest business. He convinced Sir Lawrence that a proper forest organisation was absolutely necessary for the welfare of the country. According to Brandis it was due to Richard Strachey's growing influence and to the patient perseverance and excellent tact of his successor Colonel Dickens that Sir Lawrence, as Governor General sanctioned a definite organisation of the new department in all Indian provinces.[54] The first attempt at forest legislation was the introduction of the Indian Forest Act by the British in 1865.

[54] *Ibid*, pp. 126–130.

4

Evolution of Forest Policy and Forest Acts of 1865 and 1878

The British administration had a basic mistrust for anything that was Indian, the Indian traditions or their administrative patterns. They lacked confidence in the capacity of Indians to look after themselves. This was "juxtaposed with the essential assumption that the essence of all human endeavour was realised in the West; in their rational and progressive institutions, in the scientific thought. The essence of Indian civilisation was the very opposite of the West and was expressed in the irrational practices of Hinduism and the Eastern system. Therefore the legitimisation of British rule was derived from the self-validating argument that India was essentially incapable of self governance."[1]

Much of their opinion about Indian forest administration, rather the lack of it, was reflected from time to time in their official accounts. The Forest Department was founded on the assumption that the Indians lacked sense of conservation.[2] This was the opinion held by a majority of the British administrators, though there were a few exceptions. The

---

1 Sanjay Nigam, "Disciplining and Policing the Criminals by Birth", *The Indian Economic and Social History Review*, Vol. 27, No. 2, 1990, p. 133.

2 Surprisingly, this assumption, though in a different language also runs through a recent Cambridge academic work. See Richard H. Grove, *Green Imperialism: Colonial Expansion, Tropical Island Edens and the Origins of Environmentalism, 1600–1860*, Cambridge University Press, First Indian Edition, OUP, 1995.

culmination of this feeling gave birth to the Indian Forest Act of 1865. To further consolidate and legitimise their control over forests, the British successively passed the Forest Act, VII of 1878 amended by Act V of 1890, Act XII of 1891, Act V of 1901 and Act XV of 1911.[3]

The 1865 Act empowered several local governments to declare certain areas as State Forests without in any way interfering with the rights of the people. As per the Act, government forests meant land covered with trees and brushwood or jungle. A revenue official and not a forest official determined the merits of a particular block of forest chosen for reservation.[4]

The Act empowered the local governments to draft local rules for their respective provinces for the better management and preservation of the forest, and prescribe punishments for infringement of rules. These rules, when confirmed and published were to have the force of law. Penalties could be imposed after vexatious seizures and arrests. The rules under this Act further dealt with the details of forest conservancy such as staff to be employed for administration of the forests, the procedure to be followed while creating "Reserved Forests" and "Unreserved Forests," their demarcation, etc.

However, from the point of view of the administrators the 1865 Act was imperfect and unsound in principal.[5] It professed to deal only with government forests and it even attempted to define what those forests were. According to B. H. Baden-Powell "...this was enough to ensure failure but worst of all it made no provision regarding rights of the user, it assumed that they were to be left alone, neither defined nor regulated; the rules (that imposed penalties) within the Act empowering the local government of the Provinces were not to affect

---

3 *Explanatory Notes on Forest Law*, Imperial Forest College, Dehradun, Superintendent Government Printing, Calcutta, 1913, p. 17.

4 The interests of the agriculturists presumably were amply safeguarded during the creation of the reserve forests. Berthold Ribbentrop, *Forests in British India*, Indus Publications, First Print, 1900, New Delhi, Reprint 1989, p. 15.

5 B. H. Baden-Powell, *Forest Law : A Course of Lectures on the Principles of Civil and Criminal Law and on the Law of the Forest*, London, 1893, pp. 35–36.

any rights. It is only remarkable that forest administration was not choked by such a law."[6]

The administrators criticised the law for being made and enforced locally. They found the claims for forest rights so overwhelming that real forest conservancy became practically impossible. These claims also led to a great deal of friction between the civil and Forest Officers.[7]

According to Stebbing "... the progress of forest conservancy in Bombay Presidency was hampered to some extent, as in Madras, by the inability of the responsible senior revenue officials to recognise the great importance of the new administrative work in the interests of the people. This disability affected many of the Collectors of districts and made the introduction of conservancy work quite unnecessarily difficult; for the Indian officials took their cue from their British superiors. In effect, it put back the work in some respects and in others caused it to develop on wrong lines."[8]

Besides the attitude of the revenue officials and the practical difficulties in the implementation of the Forest Act, financial difficulties and the inherent defects in the Act itself made the implementation of the Forest Act of 1865 altogether difficult. In a memorandum dated 28 February 1869, containing suggestions on certain points connected with the management of the forests in Bombay Presidency, Inspector-General of Forests Brandis stated that "In the matter of forest legislation, I would draw attention to the peculiar difficulties under which portions, at least of the Bombay forests, are placed in the matter of protection. I understand that the government forests are frequently so interlaced with private forest lands that protection is impossible without a system of strict control over all timber, wood and forest produce in transit; whether it is the produce of government or private lands or has been imported from forests beyond British territory."[9]

---

6 *Ibid*.

7 E. P. Stebbing, *The Forests of India*, Vol. II, London, 1921, p. 137.

8 *Ibid*, p. 134.

9 *Ibid*, 28 February 1869: A memorandum containing suggestions regarding the management of forests in Bombay Presidency, p. 158.

A majority of the Collectors of Madras Presidency and also the Madras Board of Revenue expressed their reservations over the Act. To quote

> "The application of the Act is quite unnecessary, in as much as the penal code already provides for offences of every description, under the heads of mischief, trespass and theft; and it is highly inexpedient to multiply legislation of a special nature and especially so, to create offences by rules which may be varied at the will of the executive.
>
> The Act will not in any way facilitate conservancy and no forest land can be protected within the scope of it, which is not absolutely the property of the government, free from private rights of every kind, for Section 2 specially enacts that its application shall not abridge, affect any existing rights."[10]

Hope, a member of the Viceregal Council, enunciated the chief deficiencies of the 1865 Bill on 6 March 1878 as follows

> "It drew the distinction between the forests which required to be closely reserved, even at the cost of more or less interference with private rights, and those which merely needed general control to prevent improvident working. It also provided no procedure for inquiring into and settling the rights, which it so vaguely saved and gave no powers for regulating the exercise of such rights without appropriating them. It obliged you in short either to take entirely or to let alone entirely. On control over private forests in the general interests of the community, it was absolutely silent. For duties on timber, even those actually levied, it gave no authority. Protection for government forests, so interlaced with private ones as to be in chronic danger of plunder, there was none. In various minor points also it was deficient."[11]

---

10 *Ibid*, Proceedings of the Board of Revenue, 10 April 1868, No. 2777, p. 14.

11 B. Ribbentrop, *Forestry in British India*, p. 97.

The Act proved to be highly defective and posed many problems for the Forest Department in the subsequent years. The defects of the Act and the details of the need to make a new Act were discussed during the series of conferences and meetings, which took place in the Forest Department until the new Forest Act of 1878 was passed.

It was Baden-Powell, who as Conservator of Forests in Punjab in 1870, held the first Forest Conference. However, the first General Conference was held in 1874 at Allahabad and the second at Simla, in 1875. The different opinions expressed by the various Forest Officers in these successive meetings led to a strong debate in which one could see three main streams of thought or opinions. Madhav Gadgil and Ramachandra Guha called them as 'annexationist,' 'pragmatic' and 'populist'.[12] The 'annexationists' were for complete control of State over the forests, the 'populists' believed that tribals and peasants who were traditionally enjoying rights over forest lands should continue to exercise these sovereign rights, while the 'pragmatic' aimed for a balance between the two, preferring selective control over forests.

In support of their belief, the advocates of 'annexationists' compared themselves with scientists, who were almost persecuted and ridiculed by the people for their discoveries. For example, Baden-Powell argued in one of his articles that "... It has often been observed that all new truth must first pass through the crucible of contradiction, doubt and even ridicule before it issues into the realms of settled conviction. Almost every branch of science, in which great discoveries have been made, has passed through such an ordeal ..."[13]

These officers believed that people were ignorant of the practical truth established by forest science. Moreover, they thought that the people were always blinded by a short-sighted idea of their own immediate interest. "All forest conservancy is therefore disliked."[14]

---

12 Madhav Gadgil and Ramachandra Guha, *The Fissured land: An Ecological History of India*, OUP, New Delhi, 1992, pp. 126–128.

13 B. H. Baden-Powell, "Popular Aspect of Forest Conservancy", *Indian Forester*, No. 1, July 1876, Calcutta Central Press, pp. 1–2.

14 *Ibid*, p. 2.

Schilich, Inspector General of Forests from 1883 to 1888, in his manual of forestry echoed similar views "History has proved that the preservation of an appropriate percentage of the area as forests cannot be left to private enterprise in India, so that forest conservancy in that country has for some time past been regarded the duty of the State."[15] The reason for this, he argued, was that it was only the State that could appreciate the importance of the indirect benefits of forests and could think in long term.[16]

Also it was presumed by them that the "... restriction is never liked at first. Even if it is such an obvious matter as prohibiting barking trees, cutting slabs and torches from living pines, removing 'humous' soil from the forest, the restriction is not popular; and hence, it is idle to speak of carrying the people with us in our efforts to conserve... Such phases are mere folly. All that in India we can hope to do is, from a stand point of necessarily superior knowledge, to ascertain facts and define our restrictions and the areas within which they are enforced with strict justice."[17]

The 'annexationists' argued that customary use of forests by the people, however widespread and enduring, was exercised only at the mercy of the Monarch. A few precedents such as Tipu Sultan's edict banning the cutting of sandalwood, were used as proof that India's rulers had reserved themselves the right of ownership over forests and forest produce.[18]

Baden-Powell made a distinction between the legal rights and prescriptive rights or privileges. He stated that there was no proof

---

15 Schlich, *A Manual of Forestry*, Vol. 1, p. 95.

16 *Ibid*, p. 89–96. Similar views were later expressed in the *General Forest Policy*, 1894 and the *Report of the Royal Commission for Agriculture*, 1928.

17 Baden-Powell, "Popular Aspect of Forest Conservancy", *Indian Forester*, No. 1, July 1876, Calcutta Central Press, p. 3.

18 Madhav Gadgil and Ramachandra Guha, This Fissured Land, OUP, Delhi, pp. 125. Also see Baillie, *Summary of Papers Relating to the Madras and Bombay Forests*, Bengal Secretariat, Military Orphan Press, Calcutta, 1851, Also see *Nature & The Orient: The Environmental History of South and Southeast Asia*, (ed.), Richard Grove, V. Damodaram, S. Sangwan, OUP, Delhi, 1998.

that the laws of Manu or Mohammadan law books ever claimed for the King or Emperor a right in all land or asserted that he was the only person entitled to be called the owner or landlord. But historical circumstances gradually produced an alteration.

> "Later conquerors developed the natural idea that they became owners of all land by their conquests. Gradually the Hindu sovereigns, chiefs, the Mohammadan princes like the Nawab of Bengal, Oudh, etc., all of whom were conquerors or had thrown off allegiance to the Mughal Emperor at Delhi and set themselves up as independent, claimed to be absolute owners of every acre of the land in their dominions. All their subjects who directly cultivated or held the land were only *rayots* or dependents; they were indeed allowed to hold their cultivated lands on certain terms and even to enjoy a hereditary right, at least to some extent, but the only real private property allowed was when the King made in perpetuity revenue free grant or recognised some noble family as holding a permanent Estate. The British Government legally, that is by the usual principle of civilised international law, succeeded to the de facto rights of all conquered or ceded States as they existed at the end of the eighteenth century and in the beginning of the nineteenth century because they were public rights established by the unquestioned power of local rulers."[19]

While distinguishing between the legal rights, prescriptive rights and concessions, Baden-Powell stated that the legal rights in their strictly legalistic interpretation were those which were granted explicitly by the State by way of grant, charter or official recognition.[20]

---

[19] B. H. Baden-Powell, *Forest Law* (A course of lectures on the principles of civil and criminal law and on the law of the forest), London, Bradbury Agnew & Company, 1893, pp. 211–12.

[20] Ibid; According to Baden-Powell such strict legal rights were almost non-existent in India. Even any individual or a village had been enjoying certain

Prescriptive rights were defined by him as "... those which might be acquired by 'prescription,' just as ownership itself could. In this case no definite origin was perhaps traceable but the user could be using the land for generations past and the user must be using them openly, peaceably and as of a right."[21]

According to him, if forest easements in India were to be established as prescriptive rights in England, two-thirds of them would not have stood the test as this required the user to show that either his or his ancestors, or the user as the holder of certain estate and those who preceded the user in that estate, possessed and continued to possess the right. Hence, he argued that prescriptive rights as they were understood in England did not exist in India. Nor did the legal rights in strict legal sense exist. But then, the British Government felt obliged to waive the technical distinctions in these definitions and left it to the officials entrusted with adjustment of questions of right in the forests to deal with what they equitably and fairly considered rights, even when some technical requisites were absent.[22]

These kinds of views reflected the notion of 'oriental despotism'. In sociology, oriental despotism is a term used to describe a despotic form of government that opposes the western tradition. Despotism is based on consent and not on force. Despot masters feed upon the servile nature of the enslaved. According to French orientalists like A. H. Anquetil Du Perron (1731–1805) the ideas of oriental despotism and the absence of private property in the east were fictions employed to justify European, particularly British intervention and conquest. After

---

rights like cutting wood or grass in the native state, it was because no one cared to interfere. It was done on sufferance with knowledge that any moment the local authorities could give away the forest land to a grantee to clear and cultivate. Such conditions prevented the growth of legal right.

21 *Ibid*, p 280.

22 *Ibid*, Baden-Powell argued that the custom "... the custom alleging that all the inhabitants of a place had a right of grazing was held bad; because there was an indefinite and fluctuating body, not that each person had a right as an ancestral one or as attached to an estate which he held." Cited in footnote no. 1 on page 280.

a stay of nearly 6 years in India, this savant concluded that the country knew only one 'absolute despot' the British Company.[23]

Poona Sarvajanik Sabha and the inhabitants of the city and camp of Poona in a memorial submitted to the government stated that "... both private grantees and village tribal communities have cherished and maintained these rights (rights over forest land) with some tenacity with which private property in land is maintained elsewhere".[24]

C. F. Amrey, one of the officers in the Forestry Conference held at Simla in 1875, preferred to call the privilege a right, if under lawful authority. He considered forest rights as alienable and unalienable. For him, public acquiring rights by prescription in public lands seemed paradoxical. He frankly stated that the shift in policy was a product of the increasing commercial value of trees. He said, "In all these lands forest rights have existed from time immemorial, we are told; but as acre after acre has been granted or settled, rights have been annihilated." He further stated "A village which pays an annual revenue of Rs. 20 to the Revenue Department will sometimes utilise forest produce to the value of Rs. 200 per annum, for which we receive nothing; and this on the assumption that they have acquired rights by prescription. And yet to this we are constantly transferring to the Revenue Department lands, which they settle free of all rights as soon as they get them, sadly in the remaining forests with the whole burden of them. If the government has not absolute control of the wastes, every acre settled involves an illegal stretch of authority."[25] To do away with this paradoxical situation he advocated that "... the right of conquest

---

[23] Cited in Andre Wink, *Land and Sovereignty in India*, Orient Longman, CUP, 1986, p. 377; In his analysis of 'Ajnapatre', a collection of Shivaji's letters to his staff, Andre Wink has shown that the 'Ajnapatre' regards the zamindars not only as office holders at the mercy of the king but as his partners and co-sharers, Ibid., p. 187; Also see Ramachandra Guha, *"An Early Environmental Debate : The Making of the 1878 Forest Act"*, *The Indian Economic and Social History Review*, Vol. 27, January–March 1990, pp. 68–69.

[24] *Memorial from Poona Sarvajanik Sabha and the inhabitants of the city and camp of Poona*, 3 March 1878, available at Gokhale Hall Library, Poona.

[25] D. Brandis and B. A. Smythies, (ed.), *Report of the Proceedings of the Forest Conference*, Simla, October 1875, Calcutta 1876, pp. 18–29.

is the strongest of all rights. It is the right against which there is no appeal."[26]

In contrast to Amrey's argument, the Collector of Nasik, E. N. B. Erskine felt that expropriation of forests would damage tribal interests. "The procedure laid down for their guidance would seem to be absurd." He wrote, "Fancy a *Bhil* or *Koli* being told to present the officer about to enquire with a 'written notice' stating the nature of his rights within a forest. I would venture to suggest that no *Bhil* or *Koli* would ever give in written notice. Could we then say that all his rights had lapsed?"[27]

The Madras Government advocated a tradition of upholding local rights and supported village interests. The Governor of Madras Presidency, Sir W Robinson, Duke of Buckingham, wrote "No legislation can be suitable to Southern India which does not in limine recognise in specific terms the ancient landmarks of village communal property, and secure to the community the income of their joint property and special interests in its administration."[28]

Sir Robinson, sharply distinguished between the State's claim to a share of the produce from cultivated land, which he acknowledged to be an immemorial one, and the absence of any such right on "communal village woodland," which he pointed out was village property, not village privilege.[29]

---

26 *Ibid*.

27 National Archives of India, Legal Department, A., Mar 1878, No. 43–142, p. 8, Appendix RRR, E.N.B. Erskine, Collector, Nasik, 11 November, 1876, Cited in Mahesh Rangarajan, *Fencing the Forests: Conservation and Ecological Change in India's Central Provinces 1860–1914*, OUP, Delhi, 1996, p. 31.

28 This was a point written in W. Robinson's minutes at the time when the Madras Government refused to have the new Forest Act (VII of 1878) applied to the province and to lengthy minutes explaining the reasons which were written in February, 1878 by the Governor Sir W. Robinson and Huddesten, members of the Council. E. P. Stebbing, *Forests of India*, Vol. III, London, 1926, p. 15.

29 Ramachandra Guha, "The Making of the 1878 Act: An Early Environment Debate", *The Indian Economic and Social History Review*, Vol. 27, No. 1, 1990, p. 65.

In their proceedings of 5 August 1871, the Madras Board of Revenue wrote as follows

> "There is scarcely a forest in the whole of the Presidency of Madras which is not within the limits of some village and there is not one in which, so far as the Board can ascertain, the state asserted any rights of property unless royalties in teak, sandalwood, cardamoms and the like, can be considered as such, untill very recently. All of them, without exception are subject to tribal or communal rights, which have existed from time immemorial and which are as difficult to define as they are necessary to the rural population... Nor can it be said that these rights are susceptible of compensation, for in innumerable cases, the right to fuel, manure and pasturage, will be as much a necessity of life to unborn generations as it is to the present... (In Madras) the forests are, and always have been, common property, no restriction except that of taxes, like the Moturpha and Pulari, was ever imposed on the people till the Forest Department was created, and such taxes no more indicate that the forests belong to the State than the collection of assessment shows that the private holdings in Malabar, Canara and Ryotwari districts belong to it."[30]

It was natural that the Madras Government rejected the rigid stand adopted by the likes of Baden-Powell. In fact the government argued "All instances of the use of the forests by the people should be taken as presumptive evidence of property therein."[31] Some of the Collectors of Madras Presidency were convinced that restrictions on pasturage and seasonal migration would lead to a decline in animal husbandry in several districts. Collectors of Kurnool and Trichinapoly pointed out that Section 26 of the Act would inflict great hardship on

---

30 D. Brandis, *Suggestions Regarding Forest Administration in Madras Presidency*, Madras, 1883, p. 20.

31 Ramachandra Guha, "The Making of the 1878 Act: An Early Environmental Debate: The Making", *The Indian Economic and Social History Review*, Vol. 27, No. 1, 1990, pp. 70–72.

hundreds of helpless women whose only means of livelihood for several months of the year consisted of gathering for sale jungle produce, while inconvenience will also be felt by active doctors, most of whose medicines were found in the forests, probably at the very season when the forest may be closed. The Collectors were of the opinion that these customary rights of use by different segments of agrarian society were both indefinite and contingent, hence codification would serve little purpose.[32]

Inter-departmental rivalry between the Revenue Department and the Forest Department also turned out to be an hindrance to the increasing control of the Forest Department over the forests. From time to time we find the difference in approach to forestry between the revenue officials and forest officials. Forest Officers by and large bothered only about growth of forests at the cost of the interests of the common people, whereas the revenue officials, for peaceful administration, had to consider the possible discontentment among the people due to the new restrictions. Hence from time to time we find accusations made by the Forest Officers against the revenue officers alleging vested interests in giving concessions to the people.[33]

Zealous Forest Officers like Stebbing have remarked that the progress of the forest conservancy in Madras Presidency was slow and the work difficult, owing to the unfortunate attitude taken up by many of the revenue officials.

The letter from Acting Collector of Cuddapah to the Secretary, Board of Revenue, on 7 April 1869 is one more instance of departmental rivalry. In the letter the Acting Collector says "I cannot see why the Forest Department should be favoured to the extent of being permitted to increase its revenue at the expense of the land revenue."[34]

Between the two extremes (annexationists and populists) was the moderate approach. There were a few Forest Officers like Brandis

---

32 *Ibid*, p. 72.

33 Early History, *Bombay Forest Commission Report*, 1887, Vol. 1; Also see D. Brandis, *Forestry in India : Origins and Early Development*, First Print 1897; foreword by Samar Singh, WWF, Nataraj Publications, 1994, p. 127.

34 E. P. Stebbing, *Forests of India*, Vol. 1, p. 100.

and Horsley, who moulded a moderate approach after taking into consideration the interests of both the villagers and the State. Gadgil and Guha named this a pragmatic approach.[35]

Unlike the annexationists, the pragmatists did not believe that the rights had to be proved in writing to give them the legal status. Brandis believed that villagers were accustomed to freely graze their cattle, cut wood etc. subject only to some restrictions from time to time. He stated "The growth of forest rights in India had been analogous to the growth of similar rights of user in Europe. There are many well known cases in which forest rights in Europe have risen out of a specific growth, and in such cases the extent of the right is construed by the terms of the grant, and is not necessarily restricted by the limitations adverted to. In most instances they had grown out of the use by the surrounding villages of the common waste or forests. Forest rights in India have had a similar origin and development as in Europe with an important difference that arbitrary dealings of the native rulers have interfered with the growth of these rights and have in many cases restricted or extinguished them."[36]

About how the native rulers treated the forest rights of the people, Brandis said "... the fact that the former rulers in many cases have extinguished such customary use of the forest in a summary manner and without compensation is hardly an argument in point, for these were the cases of might versus right. As against other individuals and communities, the customary rights to wood and pasture have as a rule been strenuously maintained."[36]

Horsley, another Forest Officer who participated in the Forest Conference, too was sympathetic towards the tribals, particularly the *Bhils* of Khandesh in Bombay Presidency. He advocated for the *Bhils* a practical monopoly in the jungle wood of Khandesh. He suggested

---

35 Madhav Gadgil and Ramachandra Guha, *The Fissured Land : An Ecological History of India*, Delhi, 1992, pp. 129–133.

36 Dietrich Brandis, *Forestry in India : Origin and Early Development*, 1897, foreword by Samar Singh, WWF, Nataraj Publications, Dehradun, 1994, pp. 133–134.

that government could establish its right over the nine most valuable species of timber all over the Satpuras. "It could also go further and close certain large inhabited tracts to the tax altogether; but outside these imperial reserves and in the parts wherever *Bhils* live, they who alone have a claim on the jungle should be allowed a monopoly of the jungle wood. They live and have always lived in the jungle, they incur constant risks from wild animals. They held these jungles as their absolute property not many years ago, and now pending their hopes for transformation into settled agriculturists, they have a distinct claim to some special rights."[37]

The Forest Department wanted its claims to be settled and defined. But Horsley says "If it were a case of antagonism – forests versus *Bhils* – and if that only half-tamed population were to get the impression that they are to be roughly treated, much of good done during the past 30 years would be lost. The case is eminently one for a compromise. Vague claims become less formidable when they are enquired into."[38]

Horsley concluded his paper with the following remarks "Every Forest Officer whose duty leads him into wild and thinly inhabited districts cannot but feel the importance of enlisting jungle tribes in his favour, and making them feel that the government, which wishes to preserve the jungle, knows that it cannot do so without the cordial cooperation of the jungle tribes."[39]

Brandis agreed with and supported the views of Horsley stating "The Forest Department has no more powerful auxiliary than the wild tribes of the Indian hills and the proof of a good officer is, he feels

---

[37] D. Brandis and B. A. Smythies (ed.), *Report of the Proceedings of Forest Conference*, Simla, October 1875, Calcutta, 1876, pp. 18–31.

[38] *Ibid*.

[39] *Ibid*; The idea of utilising wild tribes by making them foresters and forest labourers was expressed by Horsley in his paper. The same note was struck by an Indian officer, Narayan Bulal Oke. Somewhat similar views were expressed by another Indian participant in the conference, Narayan Anant Okedway. See "Report of the Proceedings of the Forest Conference of 1875: A Review", *Indian Forester*, April 1876, pp. 155–156.

sympathy with these people and endeavours to make them trust him and like him."[40]

Brandis believed that all jungle tribes had certain rights, which must be respected though they were undefined and in many cases, it was out of the question to come to a clear legal understanding about these rights. These views were very much unlike those of Baden-Powell, who believed that rights had to be legal and they had to be proved. Otherwise they were mere concessions at the discretion of the Forest Officer.

Though not being a Forest Officer himself Field Marshal Lord Roberts, V. C., in his book *Forty One Years in India*, passed somewhat objective remarks on the entire situation that prevailed in India, when people were being blasted with a series of new laws to which they were not exposed before.[41]

Roberts wrote

> "Amongst the causes which have produced discontent of later years, I would mention our forest laws and sanitary regulations, our legislation in fiscal systems – measures so necessary that no one interested in the prosperity of India could cavil at their introduction, and which are so absolutely foreign to native ideas that it is essential they should be applied with the utmost gentleness and circumspection. The proceedings and regulations of the Forest Department, desirable as they may be from financial and agricultural points of view, have provoked very great irritation in many parts of India. People who have been accustomed from time immemorial to pick up sticks and graze their cattle on forest lands cannot understand why they should now be forbidden to do so, nor can they realise the necessity for preserving the trees from the chance of being destroyed by fire or risk, to which they were frequently exposed from the native custom of making use

---

40 D. Brandis and B. A. Smythies, *Report of the Proceedings of the Forest Conference, Conference*, Simla, October 1875, Calcutta, 1876, p. 32.

41 E. P. Stebbing, *op.cit.*, Vol. 2, pp. 466–467.

of their shelter while cooking and of burning the undergrowth to enrich grazing."[42]

These remarks of the Field Marshal give us a picture of the kind of confusion that might have prevailed in the minds of the people of India, during this particular period. Nevertheless these remarks do not show the understanding for the customs and traditions of India shown by Brandis who said "There has been thoughtless talk as if the natives of India, in burning the forests and destroying them by their erratic clearings, were committing some grave offence. If the matter is carefully analysed they will be found to have the same sort of prescription which justified the commoner in the new forest to exercise his right of pasture, mast and turbary."[43]

Brandis recognised and appreciated the traditional form of forest conservation, which is sacred groves, which he said were very numerous and were found in nearly all the provinces. He also appreciated Indian rulers for their forest sense, notably the rulers of Rajasthan, who had preserved brushwood in a dry climate. The Forest Officers of the British Government will do well to emulate this example, he had thought. He was especially keen on reviving and strengthening village communal institutions. But his requests were turned down.

Finally within the official circles the balance of opinion was clearly in favour of the annexationists. Based on the memorandum prepared by Brandis in 1875 and the paper prepared by Baden-Powell the Forest Act 1878 was passed.

Thus the prolonged debate finally ended in favour of the annexationists although the Madras Government did not give up its possession until 1882, when Sir W. Robinson retired and Brandis was put in-charge of the drafting of the Madras Act.

The Forest Act 1878 provided for the following:

(1) The power to declare government lands as reserved State forests.

---

42 *Ibid.*

43 D. Brandis, *Forestry in India, 1872*, reprinted by Samar Singh, WWF, Nataraj Publications. This is from the paper read by Brandis at the meeting of the British Association, Brigton, 1872, p. 134.

(2) The power to demarcate these reserved forests and to enquire into the record and settle rights over them once and for all, as it was thought that the forests were often burdened with rights adverse to the public interest and hence, they had to be dealt with in a special way. The power to commute such rights whenever they prevented the conservation of the area as forests. The power to prevent the springing up of new rights except by government grant.

(3) Provisions for the protection and management of reserved State forests, protected forests and unclassed forests.

(4) Provisions for providing suitable punishment for breaking of the forest law. This was because forests contained a great variety of produce and acts of petty theft, trespass, mischief, which constituted forest offences, were varied. Hence it was thought that they had to be treated by a special law.

(5) The power to protect forest produce in transit to check its theft.

(6) A forest was considered to be particularly liable to accidents –some due to natural causes and others due to the carelessness or malice of men, for example fire. Hence provisions were made to prevent such accidents.

Also, the thought behind the making of the law was that of the Forest Officers who were needed not only for the sake of improving the forests but also to act as police. For it was believed that for the protection of forests a service had to be vested with certain powers of arresting offenders and obtaining help in case of need.[44]

This Forest Act was permissive throughout, that is, it was left to the local government to apply or not to apply it to any particular part of the country. It extended to all the provinces of British India with the exception of Madras, Coorg, Burma, Berar, the Hazra district, Punjab, Ajmer and Baluchistan.

With this Act provision was made for the constitution of reserved forests and protected forests. In the first draft of the Act, it was intended

---

[44] B. H. Baden-Powell, *Forest Law, "Course of Lectures on the Principles of Civil and Criminal Law and on the Law of the Forests,"* London, 1893, pp. 110–139.

to form only one class of demarcated reserve State forests and provide a limited protection to all other government forest lands until the time arrived in any particular area to decide which particular forest area should be demarcated and constituted as reserve forest. Later this was modified to also include demarcation of protected forests.[45] A reserve forest once notified was safeguarded from infringement of private rights, whereas before declaration of the protected forests, it required inquiry and recording of the nature and extent of rights of the government and private individuals.[46]

Under the 1878 Act, according to Baden-Powell, the reserved forests were the only forests par excellence because they alone were properly and fully secure. In these, all the essentials of forest constitution were observed; the boundaries were legally and strictly determined and demarcated on the ground; all rights of every kind were enquired into, recorded and authoritatively defined and provided for; all other rights not recorded were declared extinguished and no new rights were allowed to grow by prescription.

The reserved forests secured a permanent settlement whereas in protected forests existing rights were recorded but not settled. Originally it was thought that a limited protection would be sufficient for the lands, which could not at once be decided as suitable to form reserve forests. This would save them from being cleared away and devastated. At that time it was thought to be sufficient:

(a) to provide a general and inexpensive demarcation;
(b) to prohibit the conversion of the forest into cultivated land, except on special permission;
(c) to reserve a limited class of valuable trees, to which no one would have a right;
(d) to periodically close portions of the area wherever essential;
(e) to provide for making of simple rules which would aim at preventing any abusive acts and facilitate orderly working.[47]

---

45 The Indian Forest Act, 1878, deals with the reserved forests in Chapter II.

46 *Explanatory Notes on Forest Law*, Imperial Forest College, Dehradun, printed in Calcutta, 1930, pp. 97–100.

47 B. Ribbentrop, *Forestry in India*, Delhi, first print 1900, reprint 1986, p. 99.

According to Baden-Powell "In the debate in Council, as reported in the gazette, an argument was put forward that the procedure for protected forests would be found sufficient for the great bulk of forests – those wanted for the supply of ordinary material to the population; and the procedure mentioned in the Chapter II of the Act (relating to the reserved forests) would only be exceptionally needed for very valuable forests growing large timber."[48]

This view, however, was never subscribed to by the Government of India. In fact, there was a fear in the minds of Forest Officers that private rights could increase without any limit resulting in the virtual disappearance of the forests. This could also in turn extinguish the rights of the user. Baden-Powell's following comment reflects this fear regarding protected forests "To put the forests on an unstable or unsatisfactory footing in such a case, is the surest way (in the long run) to cause ultimate injury to the rights, because the forest will in time cease to satisfy them. This is only doubtful to those (if such still exist) who do not believe in the necessity of forest conservancy."[49]

Consequently protected forests were gradually converted into reserves. In 1889–1890 there were 56,000 square miles of reserves and nearly 20,000 square miles of protected forests. By the end of the century reserves extended to 81,400 square miles, while the protected forest only had an area of 8,800 square miles.

The 1878 Act, under Chapter III made provision for the constitution of village forests. The thought behind forming village forests was that village communities and private persons had a right to expect that the government should provide protection to their forest property; protection against trespass and damage and also against the accrual of new rights. The village forests were to be under the management of the government officer.[50] The surplus revenue from the village forests was to go to the village community. But the procedure

[48] B. H. Baden-Powell, *Forest Law*, London, 1893, p. 234, footnote 1.

[49] *Ibid*, p. 235.

[50] *Explanatory Notes on Forest Law*, Imperial Forest College, Dehradun, Superintendent Government Printing, Calcutta, 1913, p. 39. The Hazra regulations enabled the Deputy Commissioner under the approval of the

of the Act demanded that the forest should first be created as reserved forest. This gave rise to a suspicion among the people that the government would gradually take over the third class of forests too.[51] Hence, the concept of village forests could not be implemented successfully.

In October 1881, Brandis was deputed to Madras to study the forestry question. This was a task that Brandis wanted Baden-Powell to perform. However, finally it was Brandis himself who undertook the primary responsibility for drafting the new Madras Act. The new Madras Forest Act V was passed in 1882 and came into force from 1 January 1883. It was framed on the same general lines as India and Burma Acts but was considered by some authorities to be an improvement on several points. The procedure dealing with the protection of government land not included in the reserved forest was analogous to the chapter on protected forest in the Indian Act. This was considered to be a great improvement even over the Burma Act. But there was supposed to be one grave defect – the provision to appeal or file a suit beyond the District Forest Court, a provision that was not provided for in the other Acts.[52]

A large extent of the forests in several provinces had already been placed under some sort of preservation under the provisions of Act VII of 1865. Hence, provision was made in the new Act empowering the government to declare any such areas as reserved and protected forests, where any rights of government or private persons over any land and forest produce in such forests had been enquired into, settled

---

local government to take up any wasteland that was not reserved forest and at the disposal of the government to make it into a village forest. These regulations contain a number of rules for the management of village forests.

51 Early protests were thus directed at this government monopoly and later at the confused legal status of the third class forests. It was not clear who held the actual proprietary rights over this class of forests, the State or peasantry. See in Commentary by Ajay Rawat, B. Ribbentrop, *Forestry in British India*, Indus Publications, First Print, 1900, New Delhi, Reprint 1989, p. 23.

52 D. Brandis, *Suggestions Regarding Forest Administration in the Madras Presidency*, Madras, 1883, pp. 1–2.

or recorded in a manner which the local government thought sufficient; or would be thus enquired into, settled and recorded after the declaration if the former inquiries and settlement were considered insufficient.

Already sets of rules were promulgated by the local governments under several sections of Act VII of 1865. After the new Forest Acts for India, Burma and Madras were passed numerous sets of local rules were drafted and brought into force. These varied in different parts of the country and had been added to or modified as and when necessary.

Special legislation was required for Berar, Baluchistan, Upper Burma and Assam that were non-regulation provinces governed under similar codes compared to those, which were in force in the rest of India. The Berar Forest Law was passed in 1886. All wasteland in Berar, which had not been alienated under settlement, was the indisputable property of the State.

Regarding shifting cultivation, the 1878 Act distinctly specified that the practice of this method of cultivation at the expense of the forest conveyed no right and could be abolished at the pleasure of the government.[53]

Settlement of Rights

Settlement of rights was the most crucial aspect incorporated in the 1878 Act. As is obvious from the debate, before the passing of the Act different views prevailed regarding the rights of the people over the forests. The Forest Act of 1878 made an enquiry into the peoples' rights an essential part of their settlement program. Under the provisions of the Act, the rights of the user were subjected to certain limitations. These limitations were held necessary in order to secure the enjoyment of rights, without imperilling the maintenance of the forests which supported them. The limitations and regulations were to be maintained without any right-holder being able to claim compensation. This limitation to the rights was based on the following thought:

[53] Though hated by the forest officials, this practice of cultivation was begun to be looked at as innocuous. See for details *Report of the Royal Commission on Agriculture*, Bombay, 1928, reprint Delhi, 1983, p. 267.

(1) Possessing a right did not entitle anybody to a share in the property. No right could be exercised to such an extent as to swallow up the property or even to render it useless or valueless to the owner, though the owner was not to be deprived of its benefits.

(2) The wasteland always belonged to the governing power. But people wandered into it and took what they wanted. Since the wasteland had reduced over the period of time and population and cultivation had grown larger, it was thought necessary to have rules for limiting its use and preventing the destruction of forest.

(3) Also it was thought that the forest itself would never be secure from permanent injury, if the extent of the right was not authoritatively determined and if its exercise was not made subject to such reasonable regulations as will prevent abuses and facilitate the due conservancy of the State.[54]

Mr. Justice West, in one of the cases, had remarked that "The mere fact of having made temporary use or intending to make use actually or potentially of a particular area or parts of it, is not possession giving rise to proprietary rights."[55]

This was very much unlike the French lawyer Proudhon, a lawyer of great repute in the late eighteenth century who had successfully argued that the "usagers or right holders were co-owners in the forest, and therefore they might demand the partition of the forest so that each household or group or individual (as the case might be) should in virtue of his right, have a little bit made over absolutely to him." In accordance with this argument, a law was passed in 1892.[56]

---

54 *Standing orders of Bombay Presidency*, corrected up to 1910, Bombay, 1912.

55 This remark was made in a case of proprietary rights in Canara. B. H. Baden-Powell, *Forest Law : A Course of Lectures on the Principles of Civil and Criminal Law and on the Law of the Forest*, London, 1893, pp. 285–286.

56 *Ibid*, p. 285. This view expressed by Proudhon had come into prominence in Europe during the French Revolution.

The thinking of the likes of Mr. Justice West was very much opposed to the thinking in Madras before the Madras Act 1882 was passed. However the dominant thinking upon which the Forest Act 1878 was based was oblivious of any such ideas as the views of French lawyer Proudhon or the thoughts of Madras officials. Baden-Powell's remark on the people concerning this issue shows the extent to which their belief in conservation of forest at all costs had hardened their sensibilities. He stated "People in India often talk about the cruelty of restricting the rights of the poor people in the forests; forgetting that the right holders are mostly ignorant of and quite indifferent to the simplest requirements of forest conservancy, and are naturally unable to see anything beyond the wants of the moment, and their desire to satisfy their needs with the least possible exertion or trouble. Sooner or later, if they were allowed to do just as they please, the forest which supplies their wants would deteriorate and eventually come to be a barren waste; of this India can show many examples. But unfortunately, the process of destruction being a slow one, the authorities sometimes refuse to believe that there is any harm being done by rights."[57]

This belief in imposing limitations on the rights of users was based on the strong conviction which was provided to them in their homeland. For example, Baden-Powell quotes Cook speaking of grazing rights "Rules for the protection and limitation of the rights are among the earliest provisions of the common law," and again "It is contrary to the very essence of a right to turn out such an unlimited number of cattle by which the whole of the herbage might be consumed – such a use the law considers not as a right but a wrong." You could no more, the author goes on to say, "acquire by prescription such an unlimited right, than you could acquire a right to clip the Queen's coin."[58]

In India a much larger power of discretion was left to the Forest Officers as far as rights were concerned. The Settlement Officer had the duty of reducing vague and general claims to something definite and tangible in terms of the actual need of the right holder.

---

[57] *Ibid*, p. 306.

[58] *Ibid*, p. 306.

Yet another way to justify the limitations imposed upon rights was sought by specifying that both the easement holder (right holder) and the forest owner have their mutual rights and duties. It was believed that the one who had the right for easement could demand equitably free exercise of it and the owner could demand that he be not deprived of the use and enjoyment of it.

The scope of the easement was fixed by specifying:

(1) The subject, for example wood;
(2) The manner of obtaining it;
(3) The place of exercise of the right;
(4) The time of exercise of the right;
(5) The amount of forest produce that was to be taken;
(6) The purpose for which the produce of the right was to be put, for example for domestic purpose or for self-consumption.

Hence the sphere or extent of the right was definite. The right holder was entitled to enjoy what was consistent with the "duty", which followed from the general nature of an easement. The duty of the holder of rights in short was to spare forests. While defining rights and the extent of right, throughout the records the government seemed to have held an attitude that it was being very generous in granting rights to the people.[59]

Legally the forest owner had freedom to:

(1) Indulge in all acts that did not touch the easement in any way;
(2) Such acts that affected the easement to some extent, but were still necessary for the due maintenance of the estate and its normal working and management.

---

[59] According to Ajay Rawat, "it was acknowledged that government as a guardian of all public interests must insist on regulation of these rights so as to render possible a good management of the reserve forest in the interests of the country. However, the profit motive prevailed and still prevails. To obtain a sustained revenue from the forests, Brandis evolved a very simple method based on concepts of Dauerwold and method of Du Comprolle, " See Ajay Rawat, B. Ribbentrop, *Forestry in India*, Delhi, First print 1900, Reprint 1986, p. 19.

The second provision was clearly a loophole in the hands of the Forest Department which later became a source of discontentment among the people. Besides leaving these loopholes, provision was also made in the Act to curtail rights notably wherever such curtailment became necessary. If this curtailment amounted to serious infraction of the rights they could be got rid of entirely by compensation.

These rights were classified in the following manner:

(a) Grazing or pasture;
(b) Grass cutting;
(c) Lopping boughs and gathering leaves;
(d) Wood rights;
(e) Right to fallen leaves for litter or manure;
(f) Right to other forest produce;
(g) Hunting and fishing;
(h) Temporary and shifting cultivation.

Grazing Rights

If the right holders were allowed to graze an indefinite number of cattle in a forest and possess a large number of cattle then – (1) No regeneration or replanting would take place, if all the forest was left open. (2) The cattle would starve and the remaining forests would be ruined much quickly if part of the forest was closed. Hence it was considered best to permit only as many cattle to graze that allowed periods of rest to certain portions of the forest, without giving too small an area to graze over.

A right, unless something to the contrary was specially stated, was supposed to supply only personal and individual requirements to the holder of land in favour of which it existed.

In India, the usual method was to count the houses in the village and to consider how many head of cattle were required for each household either for ploughing or for milk. If the number varied from time to time, the average was to be taken for the number of years. It was believed that 12 years were often adopted by Indian law.

Cattle kept for commerce and not for ploughing or for household wants were as a general rule not entitled to graze, as it was against the principle that right holders should only satisfy their own individual wants, that is rights were meant for individual requirements and not for commercial use.

In short, the Indian law specified:

(1) The number of cattle allowed;
(2) The kind or description of cattle;
(3) The limits within which cattle could graze; and
(4) The grazing seasons.

There were some other particulars specified such as the presence of a responsible herdsman was essential with the cattle and not very young children were to be sent after the cattle. The roads were to be fenced in otherwise graziers were not responsible for trespass. As a rule graziers were not allowed to erect huts. However, the restrictions imposed on grazing were never appreciated by the people. Not only the people but also some officers within the government itself criticised the restrictions. For example, Commissioner H. E. M. James said, "... The large figures representing the number of cattle impounded suggest doubts whether the present system is not too violent a departure from the old established custom. The entrustment of grazing interests to a department in its very nature hostile to them can only be regarded as an experiment which should be carefully watched. It cannot be doubted that the aim of the department is to subordinate all the interests to the growth of timber. The so called forests are artificial creations of yesterday whilst as grazing grounds they have existed for generations."[60]

The Forest Administration Report for 1889–90 stated "Allusion is made to illicit grazing but customary grazing that has been enjoyed for generations should not be stigmatized as illicit grazing merely because it has now been decided to make a charge for it."[61]

Grass Cutting

Grass cutting was to be allowed in parts of the forest where grazing was likely to cause harm. However, grass cutting too was not free from

60 Appendix, *Report of the Forest Grievances Enquiry Committee*, Government of Bombay, 1927, p. 59.

61 *Forest Administration Report of the Bombay Presidency*, 1889–90.

danger. A careless cutter could destroy young seedlings and coppice shoots. Also the removal of grass injured the seedlings.

Rights to thatching grass or grass for industrial purposes were common in India. But rights to cut fodder grass were considered very rare. If grass cutting was to be a right, the amount of grass was to be fixed by the number of bundles, headloads or ponyloads.[62]

Lopping for Fodder and Gathering Leaves

Lopping of trees and bamboos was practised for *rab* cultivation. Weeds and brushwood or boughs of trees of bamboo were dried and burnt and the ashes dug into the soil. The practice was considered essential in most parts of South India and Bombay. As it was thought by the Forest Department that lopping was always injurious to a forest, it was to be restricted to species which did not give timber or trees which were to be felled in a short span of time. Rest was to be allowed to a forest after lopping. Lopping was to be permitted only after trees had reached a certain age. Supervision was required if certain kind of trees were to be lopped. Leaves were gathered for various purposes. In Berar, the gathering of teak leaves was a right. The practice of lopping trees, though disliked by the Forest Officers, was considered essential in that region by scientists.[63]

---

62 British forest officials favoured fodder cutting as against cattle grazing in the fodder reserves. They felt that cattle caused considerable damage to saplings, while fodder cutting was less harmful, though it was far cheaper to take the cattle to the fodder than to bring fodder to cattle. See Mahesh Rangarajan, *Fencing the Forests : Conservation and Ecological Change in India's Central Provinces 1860–1914*, Oxford University Press, Delhi, 1966, pp. 72–73.

63 *Bombay Forest Commission Report*, Vol. 1, 1887, p. 103. It is a fact that this practice was going on since time immemorial but there were certain norms observed by the people regarding what type of trees to be lopped, when should it be done etc., Witness No. 5, Oral Evidence, Vol. 4; Also see Indra Saldhana, "The Political Ecology of Traditional Farming Practices in Thana district of Maharashtra (India)", *The Journal of Peasant Studies*, Vol. 17, No. 3, 1990, pp. 433–443.

Wood Rights

Prior to the passing of the 1878 Act, cutting of trees was believed to have been a restricted right from ancient times. It was believed that in the past certain trees were reserved as royal trees. Rights to wood for repairing houses etc. it was believed, hardly extended to trees which were locally reserved, unless there was an express grant on the subject. Hence, the purposes for which building wood could be claimed were for repair or entire rebuilding of houses, bridges etc. But no claim could be enforced for new buildings in addition to those already on the ground. Sheds and accessory buildings necessary for the right-holders' business were included in the building right, but all other buildings were to be rebuilt as they were before. A man who had a right to a certain number of poles for a village hut could not claim beams for a larger and better house. The grant of wood was made respecting the custom and style of buildings in the locality. But the wood granted under a right was not to be sold or exchanged. To supply the material the Forest Officer received the claims of the right holders and issued a permit specifying all details and the time within which it was to be taken out of the forest.[64]

Firewood

The right for firewood was considered to have originated in the practice of taking out of the forest what was fit to burn. Hence, the rules under the Act did not classify firewood according to its size. The right to firewood was restricted to any waste and inferior stuff which was considered sufficient to serve the purpose. Permission was not granted to cut any living tree, when the people could do without it. When rights existed for green fuel, special arrangements were required to confine the work to the yearly coupe or arrangements were to be made to regulate the fellings.

---

64 *Explanatory Explanatory Notes on Forest Law*, Imperial Forest College, Dehradun, Superintendent Government Printing, Calcutta, 1913, pp. 21–23. Also see *Standing orders of Bombay Presidency*, corrected up to 1910, Bombay, 1912, pp. 191–236.

A right to cut timber trees for firewood at any stage of growth was not admitted. Badly grown timber trees and other taken out at thinnings could be given as firewood as a matter of special concession, if the wood could not be otherwise utilised.

To limit the rights in case of right to fallen and dead branches, the area over which collection was allowed had to be specified. If the right to firewood was exercised by each house in a village, the houses were counted and their requirements estimated. If a right holder enlarged his house and increased the number of hearths, he could not claim more than before.

Also the right-holder was not to cut more than his yearly quota at a time. No restriction was imposed on the mere gathering of dead and fallen branches except that climbing irons or cutting instruments were not to be used for this purpose.

Large standing, dead or fallen trees were included under the head of timber unless they were lying useless and rotting in the forest.[65]

Wood for Implements and Industries

These rights were treated as "other wood rights". Bamboos were a product demanded in large quantities and right holders were only in certain parts of the forest at a time. In this case, the number of bamboos were fixed. There was a lot of demand for wood for ploughs and other agricultural implements.

Other rights were fulfilled by wood of small dimensions and in small quantities. Their species were defined, for example boxwood for comb makers, *Khair* for shuttle of a loom etc. Also the quantity of wood to be removed and the locality of cutting was specified.

Right holders were not to make wasteful use of any produce, for example cutting young saplings to make fences. Right holders had to procure a written order specifying where the wood was to be obtained from and how many pieces were to be removed.[66]

---

65 *Ibid*.

66 *Explanatory Explanatory Notes on Forest Law*, Imperial Forest College, Dehradun, Superintendent Government Printing, Calcutta, 1913, pp. 27–31.

Right to Stumps of Trees and Roots

In India, villagers were not often able or willing to use stumps and roots for fuel, except in northwest Himalayas, as the labour and cost were too great.[67]

Rights to Decayed Leaves from the Ground

In India, this right chiefly existed in deodar and coniferous forests. It extended not only to the freshly fallen leaves, but also to those which were decayed. This right was not to be exercised:

(1) on very steep slopes;
(2) on poor stony soil;
(3) in parts where restoration and regeneration was on. However, it could be allowed when the leaves were thickest on the ground, that too under the supervision of forest guards;
(4) only the surface was to be taken and not the black soil underneath.

Rights to Other Forest Produce

These were, for example, the rights to fruits, honey, bark and leaves for tanning and fibre, limestone, gravel, resin etc. They were to be implemented in a manner such that as little damage as possible was done and wastage was avoided.[68]

---

67 *Ibid*.

68 Following is a comment from a Report on the working of Forest Department in India "The forests in India are so rich in minor products of all kinds that it is possible to refer only to those which are or are likely to be of commercial value." A large amount of this minor forest produce was exported by the British Indian Government, especially during World War I. The industries depending on these products for their raw material received a great stimulus. For further details, see *The Work of Forest Department in India*, prepared under the direction of the Inspector General of Forests to the Government of India, Forest Research Institute, Government Press, Dehradun, April 1920, pp. 24–29.

Hunting and Fishing

In India, the Forest Act did not consider this. No right was provided for. But the Indian Forest Act Sections 25 (i) and 31 (j), Madras Forest Act Sections 21 (h) and 26 (f) and Bombay Forest Act Sections 26 (g) and (h) and 33 (e) enforced rules made by the local government on the subject. The rules were made to protect forests against fire, trespass and to preserve the game itself.[69]

Temporary and Shifting Cultivation

These are methods of cultivation often used by tropical-forest root-crop farmers in various parts of the world. Areas of the forest are burned and cleared for planting; the ash provides some fertilisation, and the plot is relatively free of weeds. After several years of cultivation, fertility declines and needs increase; as the fertility declines, the cultivators move to another forested area for clearance and to repeat the same method leaving the earlier piece of land fallow.

With regard to this, two questions arise. (1) Did the practice give any claim to an occupancy, a landholder's or other proprietor's right on the soil? (2) If not, did it give right to a user to use the government forests for temporary cultivation?

The question of a right as servitude was authoritatively settled by Act V of 1890, which amended Act VII of 1878, and later amended in some particulars by Sub-Section 9A. After the amendment, the Forest Settlement Officer in India could not grant rights to shifting cultivation. He could record his opinion in the matter, but the practice of shifting cultivation in all cases was to be deemed a privilege subject to control, restriction and abolition by the local governments.[70]

Right of Way and Water Course

Right of way is the right to use a forest path for approaching one's piece of land. Right to water course is the right to use dam water in reserves from the forested areas.

---

69 *Explanatory Notes on Forest Law*, Imperial Forest College, Dehradun, Superintendent Government Printing, Calcutta, 1913.

70 *Standing orders of Bombay Presidency*, corrected up to 1910, Bombay, 1912*t*, pp. 200–207.

Under the Indian Act, the Forest Settlement Officer did not deal with right of way at all, but simply recorded it. Under the Burma Act and Madras Act, if he admitted a claim he recorded it.

The Forest Settlement Officer could, with the previous sanction of the government, stop any public or private rights of way or watercourse in a reserve, provided that instead, another right of way or watercourse equally convenient in the opinion of the local government existed or had been provided for by such Forest Settlement Officer. The reason given for this was, that at the time of settlement and first demarcation it could not be known whether such a right in such a place was injurious or not.[71]

Claims for Other Forest Rights

The Forest Settlement Officer was supposed to discover the real nature and extent of these rights. In the case of claim to pasture or forest produce, the Forest Settlement Officer could reject or admit them in whole or in part. If the claims were admitted, they had to be recorded in some definite shape, and they were to be so arranged and regulated as not to injure the forest. The right holders were responsible for any injury to the forest.

Several limitations were imposed on the nature and extent of the rights based on the mistrust for people, and their belief in people's insatiable greed for timber. In addition to that, there were several formalities to be fulfilled before rights were defined and admitted. This surely was a cumbersome task for the illiterate commoners, who were not used to the formalities. The provisions of 1878 Act required that certain conditions had to be fulfilled by the people for the admission of the nature and general features of the rights. The general features of the rights were to be obtained by a written order or a grant which was considered as the best evidence of the nature and extent of rights. Sometimes when the land settlement orders were passed they in effect amounted to grants of forest rights, but were usually couched in general terms which needed to be fairly interpreted in order to enable the rights

---

[71] *Ibid*, pp. 296–300.

to be defined. If the meaning of the term was ambiguous, the question was settled in favour of the government for the public benefit.[72]

If the right holder was permitted to sell his produce, it was to be recorded by the Forest Settlement Officer. If it was not specifically stated, it was understood that the right was for the right holder's own use and had to be measured by his own actual wants. Hence the rules of the Act required the Forest Settlement Officer to go into the intricacies like taking an average of the quantity taken by the right holder in a series of former years.

The right could not be transferred except in case of real servitude with property to which it was appending.[73]

If the rights could not be provided for under Section 14 of the Indian Forest Act, then the local governments commuted such rights either by payment to such persons a sum of money in lieu thereof or by grant of land or in such manner as thought fit. This duty was with the Forest Settlement Officer and not with the parties to the settlement. The Forest Settlement Officer had the discretionary power to consider whether the right was absolutely necessary or not.[74] If necessary, the Forest Settlement Officer could abandon the reservation of a tract if

---

[72] To establish the rights of the people they had to fulfill a number of formalities which had caused confusion in the minds of the people. See Memorial to the President and Members of the Forest Commission by the Proprietors of Freehold Villages with forests attached there to in the *talukas* of Salsette, Zilla Thana, Bombay Presidency, *Bombay Forest Commission Report*, Vol. 4, 1887, p. 37.

[73] Indian Forest Act 23, Bombay Forest Act 24 and Madras Forest Act 19. A right-holder whose right was in the form of grant had to render a service or pay a certain sum. All these provisions were at the discretion of the Forest Settlement Officer.

[74] A lot of times over enthusiastic Forest Officers were very unjust and unfair to the people needs and sentiments even in cases when the rules by themselves were not so torturous. See General Policy Guidelines, *Standing orders of Bombay Presidency*, corrected up to 1910, Bombay, 1912, pp. 317–321.

that tract was required to supply some right, which could not be commuted by purchase.

By Sections 9, Indian Forest Act, 19 Bombay Forest Act and 17 Madras Forest Act, the rights in respect of which no claim has been preferred and of the existence of which no knowledge has been acquired by inquiry were to be extinguished, unless before the publication of the notification under Section 19, Indian Forest Act, 18 Bombay Forest Act and 17 Madras Forest Act, the person claiming them has shown that he had sufficient cause for not preferring such claims before.

By Sections 16 Indian Forest Act, 14 Bombay Forest Act and 10 Madras Forest Act, claimants could appeal against the orders passed by the Settlement Officer. The time allowed for this in India was 3 months. Under Indian Forest Act 16 and Madras Forest Act 14, a Special Forest Court could be appointed in case of complications. The appeal was heard somewhere near the forest to facilitate, if necessary, local investigations.[75]

When the nature and general features of the rights had been admitted and recorded, these rights had to be provided for. In India, there were following three ways of doing it:

(1) By setting aside some forest tract of sufficient extent, in a locality reasonably convenient for the purposes of the claimants, and recording an order conferring upon them a right of pasture or to forest produce.
(2) By altering proposed boundaries of the forest. To exclude a certain portion of the area that would be sufficient for the right holders. If the people abused the right and failed because of that to supply what they wanted, they could not have any claim to return to the original forest which was freed.

[75] According to *Kesari*, people's rights were settled at the time of Forest Settlement and they were published in the gazettes. If people had any objections or clarifications to these rights, they were supposed to approach the Collector within 3 months. However, many a times people were not aware of this. In most of the cases, the 3 months period had elapsed by the time they were informed about it. Hence, it was as good as having no right to appeal against these settlements. See *Kesari*, 30 June, 1885.

(3) By recording an order, subject to regulations, the right to exercise inside the forest, the number of cattle, quantity of forest produce and the local limits of exercise etc.[76]

On the whole the rights laden with numerous formalities and intricate details were more of a curse than boon on the common people. Traditionally, they were not used to so many formalities of rules and regulations. Their dependence on forests was much less complicated. Also these rules and regulations with their intricate details were not communicated to the people properly. In the Bombay Forest Committee report and in several other documents, people are found complaining of being unaware of the new rules until they were punished for it. The innumerable restrictions on the use of forests along with numerable legal complications meant colossal supervision, reducing the Forest Officers into police officers.

Protection of Forests

The 1878 Forest Act, also perceived and made provisions for protection of forests against natural calamities by formulating the rules for maintenance of forests in a manner that they provided protection against erosion, avalanches, winds etc. With great care, the 1878 Act specified which were to be considered offences and how they were to be dealt with strictly.

Preventive measures were mentioned in Indian Forest Act 25 (c), Burma Forest Act 21 (c), Hazara Regulations 7a, which enabled a Forest Officer to prevent fire in reserved forests. By the provisions of Indian Forest Act 78, Burma Forest Act 50 and Madras Forest Act 23, government servants or persons having an interest in forests were bound to give information and help extinguish the fire, prevent a fire from getting into a forest and help to discover the offender, if any. Omission to give information of fire was punishable under Section 176, Indian Penal Code. Omission to assist in putting out a fire was punishable under Section 187, Indian Penal Code. Punishment was

---

[76] *Explanatory Notes on Forest Law*, Imperial Forest College, Dehradun, Superintendent Government Printing, Calcutta, 1913, pp. 15–31.

awarded by the India, Burma and Madras Acts for kindling of fires or setting fire to the forest or leaving a fire burning in any dangerous place. In serious cases, Indian Penal Code, 435 could be applied. Mischief by fire, intending damage or in spite of knowing that it could cause damage, could lead to a fine of Rs. 100 or upwards, and punishable imprisonment that could extend up to 7 years. It was also punishable with fine.

When a fire was caused by gross negligence the local government, according to the provisions of Indian Forest Act 25, last clause, Bombay Forest Act 56, Madras Forest Act 22 and Hazara Regulation 29, could close the forest for such time as it thought fit, suspending all rights to grazing or to forest produce.[77]

For protected forests, the local government was supposed to make rules under Indian Forest Act 31, Bombay Forest Act 33, Madras Forest Act 26 (b), to protect timber and valuable trees from injury.

By Indian Forest Act 32 (d), (e) and Bombay Forest Act 31 and Madras Forest Act 26, it was punishable to set fire to a forest or kindle fire without taking precautions to prevent its spreading and injuring timber lying in the forest or reserved trees. Those negligent and leaving the fire unextinguished were also punishable. This rule applied, whether the fire was in the forest or in its vicinity.

Forest Offences

Forest offences were divided into: (A) Offences against the forest itself; (B) Special offences; (C) Cattle trespass; (D) Offences relating to the forest produce in transit and to drift timber.

(A) Offences against the forest itself

(i) *Reserved forests* – The offences against the reserved forests were stated in the Indian Forest Act 25, Bombay Forest Act 26 and 27, and Madras Forest Act 21. The offences against the reserved. Offences

[77] Communal punishment for fire had always caused a lot of worry to the villagers. Time and again people appealed to the government to stop this punishment. For further details see *Report of the Forest Grievances Enquiry Committee*, Appendix, Government of Bombay, 1927, p. 3.

under village forests were treated as such under the Forest Act 27 which drew from the previous act.

(ii) *Protected forests* – In case of protected forests, offences were punishable under Chapter IV of the Forest Act, Section 32. But the provision applied only when the limits of the protected tract had been notified under Section 28, and nearly all offences, except that of setting fire to the forests {Section 32(a)}, depended on declarations under Section 29.

The offences punishable under Section 32 were kindling fire, injuries to reserved trees, breaches of the prohibitions and rules. The punishment amounted to imprisonment upto 6 months or fine upto Rs. 500 or both. According to the Royal Commission, "speaking broadly everything that is not permitted in a reserved forest is an offence while in a protected forest nothing is an offence that is not prohibited."[78]

(B) Special offences

Sections 61 Indian Forest Act, 60 Bombay Forest Act and 52 Madras Forest Act were made to prevent a misuse of authority by Forest Officers. Sections 62 Indian Forest Act, 61 Bombay Forest Act and 50 Madras Forest Act provided a special punishment, which was heavier than that given by the Indian Penal Code, for destroying forest boundaries. It also gave a sufficiently heavy punishment for the erasure, tampering with or forging marks in forest work. Section 434 Indian Penal Code gave one year's imprisonment or fine or both for tampering with boundary marks used in forest work.[79]

---

78 *Report of the Royal Commission on Agriculture in India*, Bombay, 1928, reprint 1983, Delhi, p. 257.

79 *Explanatory Notes on Forest Law*, Imperial Forest College, Dehradun, Superintendent Government Printing, Calcutta, 1913, p. 45.

(C) Cattle trespass

The first Cattle Trespass Act I was passed in 1871. It was later amended by Act VIII of 1890 and Act I of 1891, which authorised local governments in special localities to prescribe heavier penalties for cattle trespass than were allowed by the ordinary law. If there was a responsible person in charge of the cattle, he was bound to keep the cattle from mischief and straying. If he was guilty of negligence or misconduct, he was liable to be punished under Indian Forest Act 25 or 32, Bombay Forest Act 26 or 33 and Madras Forest Act 26 and 29. The cattle was to be impounded, when trespassing in any reserve or closed portion of protected area. This was especially specified since Cattle Trespass Act was found insufficient. According to this Act, the scale of fines depended on the value of the cattle, while under the Forest Act the extent of damage done to the forest was also taken into consideration.

The scale of fees was prescribed by Indian Forest Act 70, Bombay Forest Act 57 and Madras Forest Act 58. Under the Cattle Trespass Act, the pound was supposed to be under the control of the magistrate. The magistrate was supposed to choose the location for the pound. He was supposed to appoint the pound keeper. All police officers were bound when asked to aid the seizure in preventing resistance. The Forest Department was not credited the fees from forest pounds; they went to the magistrate of the district who had a fund, out of which he repaid the keeper and pound expenses. If a person objected to the seizure of his cattle, he could after paying fine and expenses complain to the magistrate. If forcible opposition to the seizure of cattle was made, it could be punished by Sections 24-28 of the Act. If the cattle indulged in any damage, in addition to impounding them, the owner of the damaged crop could sue for compensation. The cattle was not to be seized nor the driver was to be prosecuted when the cattle had got into the forest by accident or notwithstanding reasonable efforts to prevent it nor when they had taken refuge from a storm.[80]

The Hazara Regulation (II of 1879), Section 31 stated that if cattle strayed off a lawfully used road without negligence on the part of the owner, they were not liable to punishment.

(D) Offences relating to forest produce in transit and to drift timber

Transit of timber by land or water, was dealt by the Indian Forest Act, Chapter VIII, Bombay Forest Act, Chapter V and Madras Forest Act, Chapter V. Section 41 made provisions not only for timber but also for other forest produce.

(i) *Transit by land* – The transit of timber by land was regulated by rules made by the local government.

(1) This transit was confined only to certain routes. The confining depended on the routes available in the country. If the roads were few keeping a watch over them was easier but difficult if the roads were many.

(2) All the produce in transit was governed by issue of passes. Stations were established for examination of the passes and the produce bought out with them. There was a check on the quantum of produce specified in the pass and the amount being taken out of the forest. The omission to procure a pass was considered

---

80 *Ibid*, p. 46; This Cattle Trespass Act was found very oppressive. In the year 1908–1909, 1,90,605 cattle were impounded. Out of these, the majority of the cases were from the Central Circle where due to scarcity of rain there were no forests as such. Instead, good amount of grass grew there which was useful for cattle. However, large amount of this grassland was wrongly included in the government forests. Hence, much grassland was left for the cattle of the villagers. Due to this, the cattle strayed into government lands and the number of cattle impounded went up in this area. *Kesari*, 14 June 1914.

a forest offence and was punishable. The government made it mandatory that private forest owners too issued passes for the produce which left their forests. This measure was taken to protect the government forest from timber theft.

Application of the Penal Code for Forest Offences

The Penal Code dealt with the following:

(A) Offences directly connected with the forests or its produce, that is,

(i) *Theft* – Dishonestly taking away moveable property in the possession of the owner was dealt under Indian Penal Code 378. Criminal misappropriation, that is dishonest taking away of moveable property not in possession of the owner was dealt by the Indian Penal Code 403.

(ii) *Receiving stolen property* – Any property, the possession of which was transferred by theft, extortion or robbery or was criminally misappropriated or in respect of which a criminal breach of trust was committed, was treated as stolen property under Indian Penal Code 410. Dishonestly receiving or retaining stolen property was punishable under Indian Penal Code 411.

(iii) *Mischief* – Mischief was punishable under Section 426 where damage amounted to Rs. 50. More than this damage was covered under Section 427. This section could be used in grave cases or extensive girdling or injury of forest trees instead of Indian Forest Act 25, Bombay Forest Act 26 (b) and Madras Forest Act 21 (e). Mischief to irrigation works, roads, bridges, navigable channels etc. came under Sections 430 and 431 and mischief to cattle under Sections 428 and 429.

(iv) *Criminal trespass* – This under the Indian Penal Code 441 involved entering property in possession of another with intent to commit an offence or to intimidate, insult or annoy any person in possession of such property and of unlawfully remaining with such intent even though the original entry could be innocent. Simple trespass was punishable under the Forest Act.

(v) *Attempts to commit offences* – Generally the punishment for attempting to commit offence was half of that provided for the offence itself.

(vi) *Abetment of an offence* – This was defined by the Indian Penal Code 107. According to Section 108, whoever prior to or at the time of commission of an act does anything in order to facilitate the commission of that act successfully, was said to aid the doing of that act. By Section 109, if the offence was committed in consequence of the abetment, the punishment was the same as for the offence itself.[81]

(B) Offences indirectly connected with forest work were the following:

(a) *Unlawful assembly* – Under Section 141, unlawful assembly was five or more persons getting together with a common objective.

(i) To overawe by criminal force or show of criminal force any public servant from exercising his lawful powers.

---

81 Several instances can be cited in the "Oral Evidence" of the Bombay Forest Commission Report where people were prosecuted and imprisoned unjustly for a number of days for some trivial acts like taking away an ain tree fallen due to strong winds on their own occupied land or for removing 10 to 12 injali rafters from their occupied land. Like Chang Patil, Witness No. 13 or Pandurang Hari, a Brahmin Vakil, Witness No. 40 of Thana district, "Oral Evidence", *Bombay Forest Commission Report*, Vol. 4, 1887. The number of cases decided for the year 1879 to 1885 over cutting of tahal for *rab* in Gurucharan lands were 10. Of this in nine cases people were convicted and one was left free. See Witness No. 12, Rao Saheb Rajaram Vishwanath Gadkari, Thana district, "Oral Evidence," Bombay Forest Commission Report, Vol. I, 1887.

(ii) To resist by criminal force – the execution of the law.

(iii) To commit mischief, criminal trespass or other offences.

(iv) To deprive any person of property or of a right he may be enjoying.

(v) To compel by criminal force or show of such force to any person to do a thing he was not legally bound to do or to omit to do a thing which he was legally entitled to do.

Six months imprisonment or fine or both were to be imposed for this offence under Section 143. If the party was armed the offence was treated as aggravated.

According to the Forest Act, certain persons were bound to render assistance in certain cases. A person so bound omitting to give information was punishable under Section 176 of the Indian Penal Code or if refusing to give assistance under Section 187. A person giving information he knew was false came under Section 177. Under Section 201 any person who, knowing or believing that an offence had been committed, caused any evidence to disappear or gave false information in order to screen the offender was punishable.

In the general policy statement, government indicated that the prosecution of petty offences against the Forest Act should be discouraged as leading to a great deal of petty operation. Hence, recommendation was made as follows "His Excellency, the Governor in Council, would therefore press on all magisterial and Forest Officers, the necessity of using great caution and judgment in regard to such offences. For the benefit of the general public and the country at large, the government is bound to do all that lies in its power to recover the forests from their deteriorated condition into which, by the uncontrolled action of villagers, they have fallen throughout the presidency."[82]

---

82 *Standing orders of Bombay Presidency*, corrected up to 1910, Bombay, 1912, p. 317.

However, the provisions made by the 1878 Act indicate that every stringent legal protection was offered for the protection of forests with certain grave cases punishable under the India Penal Code. For example, the Gazetteer of the Bombay Presidency for Kolaba-Janjira district, mentions that of the number of persons convicted for all types of offences in the district, those convicted for forest offences were the highest for 3 consecutive years. The total number of people convicted for various offences like murder, dacoity, robbery, police, excise and municipal offences, hurt, criminal force and assault including grievous hurt and other offences were 1805 for the year 1900, 1708 for 1901 and 1372 for 1902. Out of these, forest offences were the highest, being 388 in 1900, 445 in 1901 and 259 in 1902.[83]

In short there were three classes of forest offences:

(i) Cases where the Forest Law could be appealed to; as with special forest offences or in cases under Section 62.

(ii) Cases where either the Forest Act or the Indian Penal Code could be followed.

(iii) Cases where the Indian Penal Code alone was applicable. And in II and III the application of Indian Penal Code could be:

(a) cases directly concerning the forests.

(b) cases indirectly concerning the forests, but connected with forest service and administrative business of the department.

For all offences where a sufficient punishment could be given under the Act and Rules, the prosecution was to be under them. In grave cases where a full penalty was called for, it was a matter for a regular criminal trial under the code, for example a serious forest fire set willfully. The anxiety to protect forests was so strong that thoughts such as follows, were expressed regarding the civil law trespass.

"If, says Addison; a man's land is not surrounded by any actual fence, the law encircles it with an imaginary enclosure to pass which is to break or enter his close. The mere act of breaking through this

---

[83] Table XVI, *Gazetteer of Bombay Presidency*, Vol. 11-B, Kolaba and Janjira, Government Press, Bombay, 1904, p. 17.

imaginary boundary constituted a cause of action, as being a violation of the right of property, although no actual damage may be done. If the entry is made after notice or warning not to trespass or is a willful or impertinent intrusion on a man's privacy, or an insulting invasion of proprietary rights, a very serious cause of action will arise and exemplary damages will be recoverable; but if there had been no insulting or willful and persevering trespass, and no actual damage has been done and no question of title is involved, the damages recoverable may be merely nominal. Every trespass upon land is in legal parlance an injury to the land, although it consists merely in the act of walking over it and no damage is done to the soil for grass."[84]

Hence, Baden-Powell says, "if simple trespass as menace to proprietary right, is actionable, so also is every act of damage whether it is cutting grass or trees, removing stones or discharging rubbish on the ground or letting water onto the land, and even if the wrong-doer has caused the damage unintentionally, he is still liable if damage actually occurs, unless the damage was done beyond his control and he could not help it."[85]

Another such example of injustice was when fire was caused – willfully or by negligence, the local government directed that in such forests, or any portion of it, the exercise of all rights of pasture or to forest produce were suspended for a period as thought fit. The justification given for such a rule was that "a chance for recovery of forests was given by such suspending of rights". It was claimed that the enhancement of punishment was not the reason behind it.[86]

In such a case Baden-Powell remarked that "... the value of such prohibition as a preventive is so great that really this consideration outweighs any hardship."[87]

---

84 Law of Torts (Caves, Fifth Edition), p. 330, cited in B. H. Baden-Powell, p. 447.

85 *Ibid*, p. 447.

86 *Report of the Forest Grievances Enquiry Committee*, Part II A, Government of Bombay, 1927, p. 29.

87 Baden-Powell, *Forest Law: A Course of Lectures on the Principles of Civil and Criminal Law and on the Law of the Forest*, London, 1893, p. 435.

In such a way stringent restrictions were imposed on the people regarding the use of the forests. People were denied their rights which had existed from times immemorial. This naturally became the cause of discontentment among the people. Doubts were expressed by the revenue officers whether they could in their position risk a great amount of popular discontent for the sake of the forests; and whether it was better to let the forest estates go altogether than have people discontented on account of them.[88] However, this discontentment went largely unacknowledged (by the government) for almost a decade until it reached it's peak in the mid-eighties leading to the appointment of the Bombay Forest Enquiry Commission.

---

88 Even in the general forest policy statement the importance of avoidance of any kind of discontentment has been emphasised. *Standing Orders of Forest Department, Standing Orders of Bombay Presidency*, corrected upto 1910, Bombay, 1912, p. 296.

5

Implementation of 1878 Act

The 1878 Act gave rise to several problems at the time of its implementation. The discontentment among people rose higher in the eighties. From the point of view of the Forest Department the 1878 Forest Act had overcome the weaknesses in the earlier 1865 Act. It was conveniently believed by Forest Officers like Baden-Powell, as seen earlier, that people did not like "restrictions." Hence discontentment was natural, though it was not to be a cause of worry. Due to this attitude of the Forest Officers, several other genuine problems were also overlooked by them for a long time.

Besides the main issue of proprietorship of forests, there were several issues which arose at the time of actual implementation of the policy rules. The new system of management required exhaustive demarcation of forest land, which required systematic surveys not only of forest resources but also of the legal rights to forest products. Simultaeneously, control over the day-to-day use of the forests had to be attempted. All this meant support of good resources and adequate well-trained staff. But the department was understaffed and its budget was to be met entirely by revenues from the collection of forest fees. By performing the function of tax collection to meet its ends, the department became an extension of the Revenue Department which was rarely popular with the rural masses.[1]

1 Richard Tucker, "Forest Management and Imperial Politics", Thana district, Bombay 1823–1887, *The Indian Economic and Social History Review*, Vol. 16, No. 3, p. 283.

The regulations of 1861 explicitly subordinated the Conservators to the Collectors who dominated the district level government. Hence, all official communication among forest officials was channelled through the Collector's Office. This gave him an effective overview of the forester's work. However, this arrangement gave the Forest Officers more than ever the identity as tax Collectors and enemies, especially in the eyes of the timber merchants.[2]

Also the new recruits in the Forest Department were not chosen on the basis of effective prior training but on vague indications of aptitude for the work. In some instances they were naturalists, in some they were sportsmen. None of them had any training in the legal complexities they would face. Neither did they have any capacity for the delicate diplomacy of negotiating with village land holders and peasants.[3]

The general policy guidelines of the government to the Forest Department demanded that, "it was essential to keep the position and authority of the Collector, as the officer responsible for the executive management and well being of his district and the head of the Forest Department therein should be safeguarded". It was further stated that government was sensitive to the zeal by which the Conservators and Forest Officers generally were actuated and of the good work which was being done by them. Unless, however, the revenue officers kept a watchful eye over all forest operations, effecting local supply and local privileges, there was evident danger of the intentions of government being frustrated and misconstrued through excess zeal for conservancy on the part of forest officials.[4]

At the local level, that is village and *taluka* level, tension developed due to the peculiar social structure. Revenue officers of the village like *Mamlatdars* and *Patils* of the village were high-caste men, often Brahmins. Belonging to the lower strata of forest service were the forest

---

2 *Ibid.*

3 *Ibid*, p. 281; Also see B. Ribbentrop, *Forestry in British India*, Indus Publications, First Print, 1900, New Delhi, Reprint 1989, p. 78.

4 *Standing Orders of Forest Department, Standing Orders of Bombay Presidency*, corrected up to 1910, Bombay, 1912, p. 297.

guards, who were required to prevent illegal use of government forest lands. For many years it was not possible to recruit high-caste men for this work as it involved arduous trips through forested tracts, which were well known only to the tribals and peasants. Hence even if the forest guards could arrest law breakers, they required the support of their social superiors to effect the arrest, which was difficult to get. A forest guard could not take action against the people who were socially superior to him even if they had committed offences. This tension between administrative and social patterns led to massive bribery of rural officials. The more the responsibilities of these rural officials increased the more widespread corruption became.[5]

Besides, the village headmen traditionally regulated the wastelands as old forest lands, which the villagers owned collectively and used primarily for firewood and grazing. However, the Forest Department was convinced that the traditional administrative pattern, barring a few exceptions, did not take care of the forests. It had a strong belief that Indians left to themselves did not have any sense of conservation.[6] Hence, at that stage they refused to allow any kind of role to village headmen in conservation of forests.

After the settling of the issue of State proprietorship over forest lands in favour of the government, certain rights and privileges were granted to the people to use the forests. However, these privileges could not prevent the discontentment among the people. One of the common complaints during this period was that, while implementing

---

5 Many witnesses in the Oral Evidence of *Bombay Commission Report* complained that they had to bribe the forest guards. For example, Witness No. 6, Thana district, Oral Evidence, Bombay Forest Commission Report, Vol. 4, 1887, p. 15; Also see Indra Saldhana, *From Shifting Cultivators to Peasants and Labourers : A Historical View*, p. 16.

6 As quoted in the beginning of this chapter, it was a general assumption that Indians had destroyed their forests very badly. For example, in forest reports of the Bombay Presidency prepared by Gibson there is ample description of the forests of the presidency which were supposedly destroyed by natives of the respective areas. *Forest Reports of Bombay Presidency*, 1856–57, 1857–58, 1858–59, 1859–60, p. 197, 200.

the forest law the interests and convenience of the local population had got unduly subordinated to other less important considerations, that provisions made by the Forest Department for local wants were insufficient and the fee levied on certain kinds of forest produce formerly obtained free was so high as to make it impossible for the poorer classes to obtain them by legitimate means. People complained that the restrictions were unnecessary and amounted to harassment. The general result was that of demoralisation.[7]

The principle objection against the rules and regulations was that they involved constant supervision and petty interference by forest subordinates. In fact senior Forest Officers acknowledged that the situation during this period had reached such a stage that the prosecutions for theft and trespass were already assuming an unenviable proportion in a Forest Officer's daily work, making him regarded as more a police officer than a friend of the people. The number of offences reported in 1878–79 were 715.[8]

The administrative formalities to be fulfilled in order to obtain daily requirements from the forests created a lot of confusion in the

---

7 *Kesari*, the Martha daily run by the militant leader Tilak stated "it has been many years since the formation of the Forest Department. However, the people have never been able to appreciate its functioning. There was always a kind of discontentment against the department and it has been increasing as days are passing. Government has started many departments so far. But this kind of discontentment is not seen against any other department... People find that the department is useless to them. Traditionally forests were the means of livelihood to the tribals and provided for people's timber and fuel requirements. However, all this has been stopped now. They feel that unless money is paid one cannot obtain anything from the jungles – since the implementation of Forest Act of 1878, the forest officials have become so anxious to show increasing revenue figures that they have started harassing people increasingly. This is the reason why people are unhappy with the department. *Kesari*, 25 May 1885.

8 Indra Saldhana, "Colonial Forest Regulations", *Nature and the Orient : Environment History of South and Southeast Asia*, (ed.), R. Grove, V. Damodaran and S. Sangwan, OUP, 1998, p. 721.

minds of the people who were not properly informed about the rules and regulations.[9]

The permit system introduced to obtain for example firewood, was not at all convenient for the people. Even for a small amount of wood they had to make three trips – one to take a permit from the *nakedar*, then to take the permit to the *naka* and then to get the firewood. After the firewood was obtained, again they had to return the permit to the *naka*. This *naka* could be as far as 2 miles from the person's village adding to the inconvenience.

Also the new system of depots created a lot of confusion. Those who habitually procured these materials by their own labour did not want to go to the depots. Many cultivators probably found it much cheaper to bribe the Forest Guards and fulfill their requirements. People wanted the passes and permit system to be restricted to the timber required for house building and repairs and not for other types of wood. Many a times, there were delays and uncertainty in obtaining free grants of wood during crisis like destruction of property by fire and other calamities. Hence many people wanted the discretionary power enjoyed formerly by the local authorities to be restored.[10]

For example, the case of free grant of wood even to the poorest like *Bhils* caused unending worry to the Forest Department in Bombay Presidency, as the Conservator of Forests objected to such grants. According to E. P. Stebbing "...the very people they objected to giving free wood, were those the department should have been making friends

---

9 *Bombay Forest Commission Report*, 1887, Bombay. In the oral evidence there were several people with similar complaints; About this complaint, *Kesari* stated that the resolutions passed by the government are not publicised the way they should be. Hence it takes very long for the information regarding the new rules to reach the people. It stated that it was no use declaring these rules in the gazettes. They should be publicised through the Talatis and Mamlatdars or by way of 'daundi'. Daundi was a traditional way of declaring new government rules, *Kesari*, 30 June, 1885.

10 Bachubin Ram Patil, Witness No. 24, Thana district, oral evidence, *Bombay Forest Commission Report*, Vol. 4, 1887.

with a view to ensuring their cooperation in forest work. And the whole sum involved was a petty Rs. 1,100 per district."[11]

Since times immemorial, people were bringing from government forests for personal consumption forest produce such as leaves, bark, medicinal herbs, roots, fruits and berries. The new regulations caused inconvenience to the residents of forest villages due to the strict reservation in forests of any products required for daily or occasional consumption. The demand for such products could not be fully met from their occupied lands.[12]

Wild tribes were to a great extent dependent for their livelihood on the money they earned by collection of minor forest produce of all kinds. The residents of villages in which there were no forests depended on labour of these wild tribes for the supply of many minor products, which were not obtainable outside forest areas in sufficient quantities. Mr. Crawley-Boevey, who served for a short time in Thana in Bombay Presidency, said "I must observe, however, that only a very insignificant proportion of the forest tribes could under any circumstance possibly be utilised by the Forest Department. Their interest as a class are diametrically opposed to the present scheme of forest conservancy. As far as my information and experience goes, no effort is at present made to enlist their sympathies or to identify their interest with those of the Forest Department. I am clearly of the opinion, that the industrial wants of the forest tribes demand the more careful consideration, for it is perfectly clear that if all principle articles of minor forest produce are severed and if at the same time all means of honest industry hitherto open to those classes are closed to them, they must necessarily either emigrate or starve."[13]

Shewan, one of the officers in Thana district of Bombay Presidency, observed "The policy of absolute prohibition of cultivation within forest limits is senselessly selfish. Cultivation could largely be

---

[11] E. P. Stebbing, *The Forests of India*, Volume III, p. 33.

[12] For further details, see Dr. Govind Gare, *Sahayadritil Adivasi*, (Marathi), Sahadhyayan Publishers, 1974; Godavari Parulekar, *Jenvha Manus Jaga Hoto*, Mouj Publishers, 19; Also see David Hardiman, *Feeding the Banias*, OUP, 1996.

[13] *Bombay Forest Commission Report*, Vol. 1, 1887, pp. 115–126.

permitted anywhere, along forest hillsides, not to mention flat tracts, without the slightest damage to the forest and to the very decided advantage of the revenue."[14]

However, a majority of the Forest Officers held an opinion similar to their colleagues like Lock and Wilkins, who did not consider it necessary or desirable to grant to the wild tribes like *Thakurs, Varlis* etc., any special privileges in the forests beyond those granted to resident forest villagers for their own use. Hence, wild tribes were one of the worst affected sections of society due to the increasing restrictions on the use of the forests.

Another officer Mackenzie, accurately represented the general feeling of the local residents on this question as follows "They saw the department cutting down trees of every kind, selling grass, preventing the cultivators from getting *rab* or wood of any real use for building or agricultural implements. They know that so far as conservancy is concerned, the department has done little but take the cutting of wood out of the people's hands into their own, and has done practically nothing in the way of replanting where they have destroyed. All this the department says the people did before, and the department called it destruction of forests; all this the department do now and they call it conservancy of forests."[15]

Another important aspect of forest conservation was restrictions on grazing area, which continued to create discontentment among all. Restrictions on grazing gave rise to major complaints such as large areas of old free pasture land getting wrongly included within forest limits. The inhabitants of villages having no forests within their limits were no longer allowed to freely graze their cattle. In case of Bombay Presidency the general complaint was that after the inclusion of Gurucharan lands in forests, the people did not know where to graze their cattle.[16]

---

14 *Ibid*, p. 125.

15 Similar views have been expressed by Christoph von Furer-Haimendorf in *Tribes of India*, Oxford University Press, Delhi, 1982; See *Bombay Forest Commission Report*, Vol. I, 1887, Vol. 3, p. 92.

16 *Ibid, Bombay Forest Commission Report*, Vol. 1, pp. 28–55.

To this the forest officials reacted by saying that the complaint had its origin in the general complaint, that villagers were no longer allowed to lop and cut trees for firewood and ash manure in the old free grazing lands included in the reserved forests.

A big controversy arose over the question of grass and grazing after the passage of the 1878 Forest Act and before the Bombay Forest Commission was formed in the Bombay Presidency. Under the pressure of famine, the Forest Department undertook the work of cutting and baling grass to be utilised as fodder for the cattle. The people did not, however, buy the cut and stacked hay, stating that the cattle would not eat it. As a result, cutting and baling was stopped and large amount of grass was allowed to remain and rot in the forests. On this, the Commissioner of Northern Division of Bombay Presidency commented "...during the previous 10 or 12 years the grasslands of the presidency had been more and more transferred to the Forest Department, until in places the department had a monopoly." The Commissioner considered this a mistake and held the opinion that large areas of grasslands should have been left for the people and not under the charge of the department.[17] The Commissioner of the Central Circle of Bombay Presidency added, that in the above situation owing to the famine thousands of cattle went to the forests, but comparatively few went again as the dry grass in the forest was not worth eating. This was yet another major cause of friction, which arose between the people and the department.[18]

In 1882, Brandis visited Bombay at the invitation of the Government of Bombay Presidency. One of the objects of his visit was to take a decision on the classification of the forest areas. He advocated the division of hill reserves into two classes – one that was to be strictly protected against unauthorised felling, grazing and fires while the

---

17 E. P. Stebbing, *The Forests of India*, Vol. 3, p. 32; Grazing continued to be a problem even in the first half of the twentieth century in Bombay Presidency. See the *Report on the work of the Provincial Fodder and Grazing Committee, Bombay Province*, 1940–41, and *The Fodder Problem in the Bombay Presidency*, 1914.

18 *Ibid.*

other, where grazing was to be permitted on payment, though burning of forests was to be prohibited. The effort of Brandis was to check the unhappy state of affairs which had developed in the presidency under which Forest Officers time and energy were occupied in impounding cattle or prosecuting offenders.

However, this did not help to remove the dissatisfaction among the people over forest administration. Matters turned so unsatisfactory, especially in the Northern Division of the Presidency, that people got the backing of influential public opinion. A public body called Thana Forest Association was formed which represented people's grievances. Poona People's Association was another public body which was dominated by Poona group, and was an active spokesman of people's grievances in Western Maharashtra.[19]

In 1881, the Poona organisation in its memorial to the government accepted, in principle, that the forests had to be more carefully controlled if they were to survive into the next century. They agreed that most severe regulations were necessary and appropriate in reserved forests but those regulations were required only there and only over the cutting of commercial timber. On all other land and for all other forest produce and grazing rights, they did not want any detailed system of permits and fees.[20]

However, despite these demands from the public the Forest Department went ahead and put into effect most of the regulations and licensing requirements for forest produce as seen earlier. In 1882, the Forest Department extended the new reservation to cover even the uncultivated village wastelands. This led to intensification of the discontent in the districts. Finally, the government in Calcutta

---

19 Richard Tucker, "Forest Management and Imperial Politics", Thana district, Bombay 1823–1887, *The Indian Economic and Social History Review*, Vol. 16, No. 3, pp. 285–286.

20 *Journal of the Poona Sarvajanik Sabha*, Vol. 3, January 1881, p. 3; The details of the People's resistance through the above mentioned public bodies like Poona People's Association, Poona Sarvajanik Sabha, Thana Forest Association, newspapers like Arunodaya, Maratha, Native Opinion of Bombay are discussed in seventh chapter "Peoples' Resistance".

announced the formation of a special commission – The Bombay Forest Enquiry Commission – to study the controversy in Thana and Kolaba districts. Its members were 4 British officials representing the Revenue and Forest Departments and 3 Indians. The Chairman of the Commission was G. V. Vidal, Collector of Thana. The other members were Rao Sahib Ramachandra, Triambak Acharya, Rao Bahadur Krishnaji Laxman Nulkar, E. C. Ozanne, C.S., Lieutenant Colonel W. Peyton, Conservator of Forests, R. C. Wroughton, Deputy Conservator and Rao Bahadur Yeshwant Moreshwar Kelkar.

The opening paragraph of the government resolution appointing the Commission was as follows:

> "The Governor in Council – wishing to secure an efficient management of forests and believing that the conservancy of forests and the maintenance of the rights of the Crown is beneficial to the interests of the people in providing for a continuous supply of timber; wishing to secure the agricultural wants of the people and the privileges they have hitherto enjoyed for the legitimate fulfillment of these wants, convinced that whatever friction has arisen in the management of forests, especially in Thana, is due to a misunderstanding which can be removed; desirous to give to all parties concerned the means of bringing their views before government – institutes a Commission."[21]

The scope of the inquiry was to be mainly directed to hearing and resolving complaints, considering the best manner in which the wants of the people could be provided. Warning was added that the Commission was not to lose sight of the fact that forest settlements were in progress, that the respective areas of reserved and protected forests were not settled completely and that the effects of the introduction of working plans on the supply of forest produce were still unknown.

The scope of the inquiry was enormous. The grievances the Commission had to enquire into did not refer to single acts in the

---

21 *Bombay Forest Commission Report*, Vol. 1, 1887, pp. 1–11.

immediate past restricting the local privilege, but to the whole policy of government as regards to trees and the uses of forests and wastelands, as pursued for a number of years in the past. The complaints were not limited to the administration of the forests under the comparatively recent Act of 1878. Some complaints are dated as far back as the year 1839 when prohibition was first introduced.[22]

The Commission's report shows following sections in which the investigation took place: grazing, timber, firewood, bamboo, exploitation, *rab*, minor forest produce, free grants of wood, wild tribes, trees in occupied lands, claims to fruit trees in waste lands, Rules under Section 41 of the Forest Act, Forest Offences and Abuses of Privileges; Forest Establishments and Trade.

One of the chief sources of complaints against the Forest Department arose from the question of grazing. The Commission acknowledged that it was indisputably the custom previous to the survey that all unappropriated and unreserved wastelands were used by the people for free grazing.

After the passing of the Forest Act 1878, certain areas of land previously assigned for free grazing along with the unassigned wastelands, were included into the new Imperial Reserves. Earlier when the free grazing lands were converted into village forests by the Demarcating Officer, the area for free grazing was in no way restricted. But the areas of free grazing in reserved forests became no longer free, and were subsequently sold by auction by the Collector's orders. Until the passing of the Forest Act, this area was comparatively small and hardly pressed on the people. The action taken in 1879 as regards the inclusion of wastelands in reserved and protected forests, consequent on the passing of the Forest Act very radically changed the arrangements of the survey, as regards free grazing. While framing the lists all lands previously demarcated or set apart as imperial and Village Reserves were included as Reserved Forests. All other available waste areas were provisionally constituted as Protected Forests, leaving their ultimate constitution to be settled after inquiry.

---

22 *Ibid*, p. 86.

The total area assigned for free grazing by the survey in Thana amounted to 6,60,485 acres. Of this 4,01,566 acres were included in and notified as Reserved and Protected Forests under Sections 34, 19 and 28 of the Forest Act. The Commission acknowledged that due to the above arrangement severe hardship had been unwittingly imposed on the villagers. The notification of reserves gave the Forest Officer the right to prohibit free grazing. In the process the number of prosecutions increased and the forest staff became little more than police officers and a situation arose giving rise to the appointing of the Commission.[23]

The recommendations of the Commission were as follows:

Residents of forest villages were allowed to graze their cattle and cut grass at no cost in the unclosed portion of the forest block. The grazing rights of each village depended on its cultivated area. How much former free grazing land was contributed by each village was not taken into consideration in determining the grazing rights. All residents of a village would have these rights regardless of whether the land taken from the village was grazing land or not. Anyone cultivating land in the village, regardless of whether they were a resident or not, were given grazing privileges. Professional graziers resident or not would be excluded from this privilege.

Allegations were made by the people at the time of the inquiry that there was a decrease of cattle in Thana in the immediate past, that is after the passing of the Forest Act. The response of officials to this was that it was certainly not due to the consequence of stricter forest conservancy, as the interference with the former customs had not started by then. They suggested that the decrease in the number of cattle was probably due to the export of grass from occupied lands.

No limit was to be imposed on the number of cattle to be allowed for free grazing in Thana district. In other places a limit was to be imposed in proportion to the area cultivated by each individual. In case of non-forest villages, some officials felt that in olden days sometimes a village was split into two or three villages. In this split one village lost its forest to the other village. Such a village should

---

[23] *Ibid*, pp. 28–55.

have a claim to the other's forests. A second view was that sometimes a non-forest village was attached to the forest of the other village and the forest was ample to take care of both. In such cases, there was no problem in allowing free grazing to the non-forest villagers.

However, the view which dominated the final recommendation was that unless some fee was charged from the non-resident villagers, the resident villagers would lose any incentive to protect their own forests. Hence the Commission recommended moderate fees for grazing of the cattle belonging to non-forest villages.[24]

Rab cultivation was another major question before the Commission. The practice of *rab* meant the practice of burning layers of cow-dung, tree loppings, shrubs, leaves, grass and clay in various combinations. The loppings of trees and shrubs used for *rab* cultivation were termed as *tahal* or *shindad*. In northern Konkan in Bombay Presidency, this practice was followed regularly to cultivate rice. *Nagli* (*Eleusine coracana*) and *Vari* (*Punicum miliare*) were also grown in rotation with other crops or with intervening fallows by the *rab* method.

In 1882, a Committee of Revenue and Forest Officers, which assembled at Thana to consider various forest questions pressing for settlement, considered that loppings were not absolutely necessary as manure for nurseries. They were of the opinion that cow-dung was preferred by the better and more industrious cultivators. This Committee recommended that with the exception of loppings permission could be accorded to collect the other materials from the unclosed portions of the forests. If they were insufficient, the department was to provide depots from which the cultivators could make up their deficiencies on payment of moderate fees. They stated that proper conservancy of forests was impossible, if the *tahal* system was to continue. The recommendations of the Committee were approved by the government in the following words "The lopping of trees cannot be allowed, nor is there any reason why the inhabitants of villages in which no land are included in forests should be permitted to take *rab* materials from the government forests. The cultivators of this latter class of villages must procure their *rab* from their own Varkas

---

24 *Ibid*, pp. 53–56.

or other lands; they clearly have no claim to collect it from government forests."[25]

In sharp contrast to this, the findings of the Bombay Forest Commission were as follows:

(i) *Rab* as practised in the northern Konkan was good farming.

(ii) That there appears no reasonable prospect of replacing *rab* by a cheaper artificial manure or by any method of cultivation customary in other parts of India or elsewhere.

(iii) Without first exhausting the available supply of cow-dung, shrubs and brushwood for the principal layer, the use of loppings from valuable timber trees for *rab* was unwise; that therefore in the best interests of the *rayot* and forest conservancy the State should, as far as practicable, insist on the full utilisation of cow-dung, shrubs and brushwood before taking recourse to trees which were valuable for timber and firewood.[26]

The Commission expressed the opinion that the supply of grass, leaves and loppings was not in excess of the requirements of the *rab* cultivation. The *rab* cultivation was confined to the better class of low-lying lands, as opposed to the Varkas or hill lands where other forms of cultivation were prevalent. They agreed that although lopping valuable trees was detrimental to forest conservancy it was perhaps a fortunate circumstance that only those forest lands which were situated within a very short distance of the cultivated fields could be profitably

---

[25] E. P. Stebbing, *The Forests of India*, *Varkas* land was that land which was not of superior quality. Only coarse grains like *nagli and vari* could be grown on this kind of land, pp. 43–44.

[26] Ozzone, an agricultural scientist from Cuencester Agricultural College stated that "The opinion held by the natives that the chief use of *rab* is to roast the soil rather than to act as manure is substantially correct, the object of roasting being apparently to destroy the germs of vegetation which would choke the young plants." See "Annual Report of the Director of Agriculture, Bombay, 1883–84", *Indian Forester*, June 1885, pp. 20–21.

exploited for *rab* material. The average weight of this material to produce seedlings for an acre of rice was after experimentation found to be around four tonnes and therefore it did not pay to carry *rab* for more than a few miles.[27]

The Forest Officers were not in agreement with these findings of the Commission. Wroughton wrote a Memorandum of Dissent with the report as follows "Whereas an Indian Commissioner dissented these findings on the ground that the recommendations of the Commission did not go far enough in the interests of the cultivators... Finally, permission was granted to continue lopping of trees. This was done to compensate for the shortage of *rab*. It was a temporary privilege granted. Final decision regarding this was to be taken later. This was granted in addition to the privilege removing grass, leaves etc. from the same area."[28]

According to Stebbing, the recommendations made by this Commission were an essential compromise considering difficulties like the inevitability of considerable friction with the masses (for the question was very freely discussed in the press) if the *rab* practice was put to an end. It was the firm conviction of some of the members of the Commission that *rab* was an absolute necessity to rice cultivation.[29]

As regards to trees on occupied lands, the right to which was vested with the government, the commission proposed that in the interests of *rab* cultivation, preservation of such trees be encouraged to provide loppings on a permanent basis. The government waives its royalty rights and gives over the rights to the cultivators of such lands with or without payment, with the stipulation that the trees were not to be sold to contractors but should be used solely for agricultural

---

27 Indra Saldhana, "The Political Ecology of Traditional Farming Practices in Thana District", Maharashtra. This paper was presented at the workshop on traditional technology in Indian agriculture, New Delhi, 6–8 March 1989, pp. 434–441.

28 E. P. Stebbing, *The Forests of India* Vol. 3, p. 46; Also see *Bombay Forest Commission Report*, Vol. I, 1887, pp. 209–212.

29 *Ibid*, p. 44.

purposes. The Commissioners expressed their opinion that once the people owned the proprietary right to trees they would be interested in their protection and maintenance.

On other forest produce, like firewood and wood for building and implements, the Commission's inquiry came to the conclusion that:

(i) A certain proportion of the population of Konkan districts required firewood but could not pay for it. However strict the enforcement of forest regulations, taking into consideration the former customs, no scheme would be practicable.

(ii) There were certain sections of the society whose livelihood depended on the collection of firewood. They would collect firewood from the forests and sell it to the villagers. This need was recognised and it was decided that this need should be satisfied. Here, headload is a unit to measure quantity of wood. (It is the quantity of wood that can be carried on the head of a person.)

(iii) At the same time, for proper preservation of forests it was considered absolutely necessary to prevent all irregular exploitation of either deadwood, livelihood timber or firewood.[30]

Consequently, the Commission recommended that:

(i) Special cuttings could be made for local supply of timber and firewood.

(ii) All small branchwood was permitted to be removed free by head loads. Free permission was to be given for this. The privilege was not limited to the members of the so-called tribes, as the Commission thought that no local resident would be interested in the privilege of carrying a head load of firewood, unless he was compelled to do so by extreme poverty or if he could not afford to pay someone else for the trouble.

(iii) Further reservation, if necessary, of branchwood in firewood contract for export was recommended. All branch and small

---

[30] *Bombay Forest Commission Report*, Vol. 1, 1887, p. 68.

wood of two inches or less in diameter could be reserved for local use. Wood of this description was not likely to be much in demand for export and there was little chance of traders, employing wild tribes or other labourers to collect them.

(iv) The prevailing privilege of collecting deadwood was to cease, except in very special cases and under strict supervision. The Commissioners believed that the supply of branchwood obtainable from special cuttings, supplemented if necessary by the supply available from contract in departmental fellings for trade purposes, would amply meet the firewood demands of the poorer classes.

(v) Passes for firewood bought at forest fellings were to be given to cover transit from coupes to destination within the district.

(vi) Rates were fixed for building wood for residents of forest villages and other neighbouring villages. For example, teak of first class quality cost from Re 1 to Rs. 1.25 per pole.

(vii) Wood suitable for agricultural purposes was to be disposed of in the same way, at rates to be fixed on the same principle as the prices previously charged, that is the agricultural prices.

(viii) Extra charge was to be collected from inhabitants of villages lying outside the forest circle.

A few of the officers objected to these recommendations. One of them, Atkins, thought that the scheme would be fatal to the retail trade, which instead could be encouraged by giving a practical monopoly of the local supply in tracts where demand was great. Another officer, Shuttleworth, felt that the scheme would involve a large sacrifice of public revenue.

The response of the Commission to this was that there was little chance of well-to-do classes getting their firewood for nothing under the cover of this privilege, as the wood made available was not of the kind they habitually consumed. The Commissioners thought it would be cheaper to buy than go and cut it for those who did not live close to the forests.[31]

---

31 *Bombay Forest Commission Report*, Vol. I, 1887, pp. 68–72.

On these recommendations, the Thana Association expressed doubts whether the Forest Department would be able to carry out the tasks entrusted to it. The association thought that the scheme was open to grave abuse. Based on past experience the Commission agreed to the objection. However, it thought that considering the merits of the scheme all that the Forest Department had to do was to ensure that the local demand was fully and conveniently met, and leave the distribution to the people themselves depending on their own wants and convenience.[32]

In case of minor produce, with the exception of *hirda* and *beheda* nuts and *mowra* flowers, villagers were allowed to collect free of charge for use, barter or sale all unreserved and unfarmed minor produce such as fruits, leaves, bark herbs and roots for medicinal or religious purposes. The Commission believed that the inhabitants of forest villages would have no good grounds of complaint, if the list of strictly reserved products was limited to articles which had considerable export demand, and for which the local demand if any could be satisfied from other than forest lands.[33]

On this recommendation, E. P. Stebbing later commented in his book as follows "This suggestion affords sufficient evidence of the ignorance existing at the period on the subject of the enormous potential value of the minor products of the Indian forests."[33]

There were conflicting opinions among the officers regarding privileges of wild tribes. For example, Wilkins did not consider it necessary or at all desirable to grant to *Thakurs, Varlis, Katkaris* and other so-called wild tribes any special privileges in the forests beyond those granted to resident forest villagers for their own use. He thought that the withdrawal of the privilege of retail sale in forest produce would not cause hardship to the tribals, as there was ample legitimate work available during different forest operations in the district.

However, the statement of the Committee of 1863, quoted in the Bombay Forest Commission Report, 1887, on wild tribes stated "It is

---

32 *Ibid*, pp. 72.

33 *Ibid*, pp. 104–108; also see, E.P. Stebbing, *Forests of India, op.cit.*, Vol. III, pp. 43–47.

hoped that these people will find employment under the contractors cutting for the department, but the employment which the department can furnish is limited to the forest which can be worked with profit, whilst the aggregate number of these tribes scattered over the forest is considerable. Moreover, their employment appears after all to be the option of contractors. Should these forest tribes take up the idea that they are harshly treated by the operations of the new rules, they are not unlikely to retaliate by burning or destroying the forest, and I do not think they could easily be prevented. They have spent their lives in these forests, are intimately acquainted with them and the whole of the Forest Department backed by the Collectors will not be able to keep them out of the reserved tracts if they have a mind to go on. Were these people a more civilised and more influential class of society, they would speedily remonstrate against measures likely to interfere with their modes of life, or get friends to do so for them. But they are very needy and ignorant and unable to comprehend the drift of measures, until practically brought home to them when this remonstrance is likely to assume a shape which will benefit neither party. A timely consideration of the claims of these people will do more for the success of the new rules than any number of repressive measures."[34]

The general conclusion drawn by the Commission was that the tribals would have no interest in preservation of the forests, if the regulations made it impossible for them to earn a sufficient livelihood therefrom. The Commission concluded that as long as these tribes were practically dependent on the forests for their living, they were to be considered as first claimants on the forests.

The Commission proposed not to grant any privileges to the wild tribes which would not be shared by all the poorer inhabitants of forest villages. These privileges, it thought, were necessary to enable the classes who were dependent on forests to earn a sufficient living and to prevent their becoming a source of anxiety in the future. The arrangements recommended by the Commission were to enable the wild tribes, as well as any other poor inhabitants of forest villages,

34 *Bombay Forest Commission Report*, Vol. 1, 1887, pp. 115–126.

driven by choice or necessity to earn their living to take free of charge head loads of firewood and unreserved and unfarmed minor forest produce including *karvi* for sale or barter. Wild tribes had the privilege of cutting and removing grass for sale from the forests free of charge. But it was a nominal privilege as there was no demand for head loads of forest grass outside. Hence it was to be discontinued. Another privilege which the wild tribes had and which was not enjoyed by forest villagers was taking free of charge wood of inferior kinds for the construction and repair of huts. The thinking behind this was that any regulation to prevent them from doing so would be difficult to enforce and so long as the privilege was limited to the material they actually required major injury to the forest would be apprehended.

Practical distinction drawn between the wild tribes and poor forest villagers, was only regarding the above mentioned privilege and the monopoly of all labour in connection with departmental forest operations.[35]

The question of right to trees in occupied lands in Thana created a major problem for the Forest Department. Great uncertainty prevailed in the minds of the people and of the Forest Officers regarding the scope and intentions of the numerous conflicting orders passed on the subject. There was very little direct evidence of the extent to which the former government exercised seigniorial or other rights over the trees on occupied lands. It appeared from the evidence available at *Chitnis Daftar*, Poona, that though 6 types of trees were reserved as royalties under the native rule, the rule was not applicable on private lands.[36]

Finally the Commission recommended that no one was on any account to be allowed to cut without an order from government any teak post, rafter etc. The *Patils, Fouzdars, Karbharis* and *Balutedars* such as Mahars enforced the order. In spite of this condition, if any person cut any teak tree owing to the want of proper management the Watan of the person, in whose village such cutting took place or probability of that cutting had existed, was to be attached. A stringent order was

---

[35] *Ibid*, pp. 125–126.

[36] *Chitnis Daftar*, Rumal No. 7663, Bundle No. 14. Available at *Peshwa Daftar*, Poona.

to be issued restricting the cutting of teak without permission from the government forests.[37]

The Commission's exhaustive survey of opinion, official and non-official was completed in 1887. The Commission had worked for 2 years and came out with systematic and far-reaching strategies. In their own way, the officers on the Commission tried to combine the two distinct and diametrically opposite objectives – that is preserving the forests in the long run and meeting the immediate subsistence needs of the villagers. They called for rapid conversion of government lands into reserved forests. By then it was established that right of full control over unoccupied village lands was vested in the government. These lands, according to the Commission's recommendations, were to be utilised to meet the subsistence needs of people but only to as little an extent as possible. Instead, the subsistence needs were to be met from private lands to the maximum extent.

The Commission criticised the Forest Department for high-handedly and suddenly imposing restrictions on subsistence use, though in principle it agreed with the view on take over of village common lands and wastelands. But the Commission acknowledged that with the imposition of new rules and regulations the villages were suddenly deprived of large portions of their subsistence base. It was for introducing the new system in village wastelands but wanted it to be done patiently and deliberately, fully acknowledging the villagers' minimum needs. The Commission wanted the rules regarding privileges to be simple and consistent enough for the villagers to comprehend and also simple enough to administer so that friction between officials and villagers could be minimised.[38]

The report of the Commission was submitted to the government in 1887. However, it did not take any action for 2 years. A resolution was thereafter passed keeping the report in view. The *Kesari* described this resolution as unjust.[39]

---

37 *Bombay Forest Commission Report*, Vol. 1, 1887, pp. 127–169.

38 *Ibid*, Vol. 1–4, 1887.

39 *Kesari*, 10 September 1889.

As the government largely ignored the recommendations of the Commission, the discontentment among the people continued to increase. This was obvious from the fact that forest offences continued to increase. For example, 571 offences were reported in 1889–90 in Thana district which increased to 1325 in 1892–93.[40] Petty thefts continued to take place. Forest fire was another matter of major concern for the forest officials. It was observed that a large number of forest fires were due to the irritation felt by the people at the strict enforcement of forest regulations.[41] The Forest Department's strategy of increasing vigilance and threats of punitive action including withdrawal of the privilege were only partially effective in controlling fires.[42]

The passing of the resolution by the government to discontinue the privilege of collecting deadwood in the forests in one *taluka* in 1894 and later its enforcement throughout the district led to discontentment among the people causing great unrest throughout the district. A series of bad seasons, high prices of grains and the general state of poverty had reduced the poor to despair. The new proclamation prohibiting forest tribes from taking head loads and cart loads deprived them of the only means of making a livelihood. Little wonder that the greatest unrest prevailed throughout the district and the people of each *taluka* held mass meetings and memorialised the government to have the obnoxious notification withdrawn.[43]

---

40 *Forest Administration Report of the Bombay Presidency including Sindh*, 1889–90 and 1892–93, Bombay.

41 *Revenue Department Files*, Vol. 48, 1894, pp. 24 and Vol. 151, 1888, Maharashtra State Archives, Bombay, pp. 65.

42 Indra Saldhana, Colonial Forest Regulations, *Nature & The Orient: The Environmental History of South and Southeast Asia*, (ed.), Richard Grove, V. Damodaram, S. Sangwan, OUP, Delhi, 1998, pp. 721.

43 As a reaction to this the privilege hitherto granted by the government in the reserved forests, namely to tribals to collect firewood free for sale and to others on payment of a fee, were withdrawn. Anyone seen fetching loads of firewood from forests after September was liable to a punishment of maximum imprisonment for six months, or a fine of Rs. 500, or both. The justification offered was that the privilege had been abused, resulting in injury to the forests, and the tribes had 'turned a concession into a trade', *Ibid*, pp. 722–723.

People held meetings all over the district, contractors of the Forest Department were boycotted and it was decided not to extend food and lodging facilities to the forest guards. A meeting of 500 villagers held at Murbad *taluka* decided to beat up the forest guards leave alone working for wages under them. In early October, the tribals actually carried out their threat of bringing firewood from government forests by defying the forest keeper and sold the same in Bassein *taluka*. The agitation continued till December 1896. This lead to a strong police force being posted in various places like Mahim and Bassein and the situation was gradually brought under control. However, as a conciliatory measure the government passed a resolution dated 22 December 1896 suspending the prohibition on headloads.[44] The British officials suspected that politically influential Brahmins on the Municipal and *Taluka* Board were in support of the forest agitation. As the agitation continued, their suspicions only became stronger.[45]

By the end of the nineteenth century, though amid great discontent among the people, the fact that forests belonged to the State was established all over India and the forests were demarcated, surveyed and mapped before being entrusted to the care of the Forest Department. After settling the forests, thus a need was felt to lay down certain general principles on which they were to be managed. At this stage the Government of India invited Dr. J. A. Voelcker to examine the state of agriculture in the country and suggest how it could be improved.

Dr. Voelcker submitted his *Report on the Improvement of Indian Agriculture* in 1893. In Chapter VIII of this monumental report, he discussed at length the conditions of the forests and stressed the need for formulating a forest policy. On the basis of his recommendations,

---

44 *Ibid*, pp. 723–725.

45 The connection with the 'notorious Tilak' and the Sarvajanik Sabha, which was said to be carrying out an against payment of revenue assessment, was noted with alarm, Persons close to Tilak were reported to have held meetings in the district, to maintain a general agitation against the revision survey, the payment of assessment, the forest restrictions and many other things, *Ibid*, p. 725; Also see *Revenue Department File*, Vol. 30, 1898, p. 69.

the Government of India issued Resolution No. 22 F dated 19 October 1894, declaring its forest policy. This was a landmark in the history of forest management in India.[46]

Following were the main features of this 1894 policy:

(i) The sole objective of management of forests was to be directed at promoting the general well-being of the country.

(ii) The maintenance of adequate forests was aimed primarily for the preservation of the climatic and physical conditions of the country and secondly to fulfill the needs of the people.

These were subject to the following conditions:

(a) Permanent cultivation was to come before forestry.

(b) The satisfaction of the needs of the local population at non-competitive rates, if not free, was to override all considerations of revenue.[47]

After the fulfillment of the above conditions, the realisation of maximum revenue was to be the guiding factor. The government-owned forests were classified with reference to their primary functions as follows:

(i) forests, the preservation of which was essential on climatic or physical grounds;

(ii) forests that afforded a supply of valuable timbers for commercial purposes;

(iii) minor forests, and

(iv) pasture lands.

---

46 Circular No. 22 F Dated 19 October 1894, A appendix V, Forest Policy Report of the Royal Commission on Agriculture in India, Bombay, 1928, reprint New Delhi, 1983; *Hundred Years of Indian Forestry*, 1861–1961, Chapter III, Forest Policy.

47 The book, *The Work of the Forest Department in India*, while describing the 1894 Forest Policy states that "Forestry has been termed the handmaid of agriculture, and nowhere does this apply with greater force than in India, which is essentially an agricultural country." See *The Work of the Forest Department in India*, prepared under the direction of the Inspector General of Forests to the Government of India, Imperial Forest Research Institute, Government Press, Dehradun, pp. 4–10.

The above classification was not intended to be rigid and a particular forest could fulfill more than one function. Though this policy was partially a response to the serious discontent among the agricultural classes caused by strict forest administration it was based on the Principle of State Monopoly of Forests. While the policy seemed to have clear bias towards serving agricultural interests more directly, it cautioned that the needs of the agricultural population were to be met only "to the utmost point that was consistent with Imperial interests."[48]

According to the Forests Officers, the main weakness of this policy was that it covered all the main issues but did not always give clear indications of choice between alternatives, nor did it lay down an ideal or appropriate scale of operations.[49]

However, this policy remained valid throughout the British rule in India, though it was much under discussion and not always followed, particularly after forests had become a transferred subject. Meanwhile, further steps were taken for technical education and training of persons to fill the controlling and executive branches of the forest service. The early appointments were made with men selected from the army, police and other branches of public services.[50] In 1885, training commenced

---

48 Madhav Gadgil and Ramachandra Guha, *The Fissured Land: An Ecological History of India*, OUP, Delhi 1992, p. 135. The Forest Policy clearly stated that the claims of cultivation are stronger when the claims of forest preservation. Accordingly, whenever an effective demand for cultivable land exists and can only be supplied from forest areas, the land should ordinarily be relinquished without hesitation.

49 *Hundred years of Indian Forestry*, 1861–1961, Government Press, Dehradun, Chapter III.

50 In June 1867, the Government of India had issued orders of a general character on the subject of promoting native subordinates in the Forest Department. With the employment of carefully trained native subordinates and officers in each Province, it was expected wood manage the forest with more thoroughness and economy. Employment of European officers for this purpose was inexpedient on other grounds and impracticable score of expense. *Appendices to the Report of the Public Service Commission, 1886–1887; Appendix No. 0.5 (Notes of Sub-committee Relating to Forestry)*, Government Printing Press, Calcutta, 1888, pp. 430–435.

at Cooper's Hill in England, as difficulty was experienced in training probationers on the continent. In 1878, a Forest School was inaugurated at Dehradun. The school turned out trained forest rangers for all the provinces. A Provincial Forest Service was inaugurated in 1891 with a view to recruit a suitable cadre of Forest Officers in India itself.

The same period also saw considerable progress in the drawing up of forest working plans. Brandis (1865–1883), Schlich (1883–1885) and Ribbentrop (1885–1900) were the three Inspector Generals of Forests who shaped the forest administration in India during that period.

6

Forest Policy During 1900–1950

In 1906 the Imperial Forest Research Institute was established with 6 officers – the Silviculturist, Superintendent of Forest Working Plans, Forest Zoologist, Forest Botanist, Forest Chemist and Forest Economist. According to Richard Tucker, the pattern of foreign training followed by British and urban Indian elite's dislike of wandering on remote mountains for several months at a time resulted into the fact that even in 1906 only two Indians had entered training for the professional ranks.[1] It was then suggested that as in other departments certain number of direct appointments should be reserved for the natives of India, trained in India and those who had not served in the inferior grades. The experience of the Government of India was that "the life of the junior Forest Officer, it was observed, was a very trying one, involving considerable amount of exposure and fatigue if he really

---

1 The forest service in India remained centralised the in structure until 1935, it did not become either more responsive or more subordinate to local interests and it remained the opponent of nationalist pressures. Its professional staff where all Europeans, trained in France until 1885, then in England until 1920 and only thereafter at the Forestry Institute which was established at Dehradun in the Himalayan foothills in 1906. Richard Tucker, "Forest Management and Imperial Politics", Thana district, Bombay 1823–1887, *The Indian Economic and Social History Review*, Vol. 16, No. 3, p. 297.

does his duty in forests. Native gentlemen in 9 cases out of 10 found the work quite uncongenial to their tastes and they would at best perform the duties in a perfunctory manner."[2] This route towards making the forest management localised was not promising and it led to further widening of the gap between the British forest management and the subsistence practices of the villagers.

The attitude of the British government towards the villages and their understanding of the subsistence agricultural practices did not change all through their administration in India. The Imperial Gazetteer of Bombay in 1909, is an example where it was stated that in spite of the great care taken by the Forest Department to control forest operations in the interests of the people, these operations did not become popular. The reason behind this was perceived as inability of the people to comprehend the necessity of foresight in forest utilisation. As mentioned earlier the basic assumption of British administrators was that the colonised people lacked any sense of conservation. In this gazetteer it was stated that as a rule the peasants were wasteful. They did not hesitate to burn valuable forest for the sake of a temporary supply of green fodder or to lop and fell trees to provide manure for their crops without thought, as to whether the supply of forest produce will continue to meet the needs of his successors.[3]

Similar attitude was reflected in the report of the Fodder Committee which was supposed to solve the fodder problem in the Presidency, especially during the famine period. The Committee discussed the different ways in which the *rayot* could economise on the use of fodder.[4] Whereas the increasing demand for timber and forest produce brought about by World War I conditions were appreciated

---

2 *Appendices to the Report of Public Service Commission, 1886-87*, Government Press, 1889, pp. 435–437.

3 *Imperial Gazetteer of India, Bombay*, 1909, Vol. I, pp. 64–65.

4 The basic assumption was that the *rayots* or peasants were wasteful in the agricultural operations and how to change their habits; these were the problems which were given more importance by this Committee. The fact that the main cause, which was the commercial orientation given to agriculture by the British government, had reduced the area which grew fodder in the past. This reduction in the acreage under fodder combined

for having stimulated organisational changes and forest development in the Indian subcontinent.

The idea of forests becoming a transferred department in all provinces had given rise to the possibility of the Government of India, having to do less and less with forest management. In 1919 under the reforms of Morley-Minto, forests were not necessarily made a transferred subject. In Bombay and Burma the Forest Departments were subsequently transferred in 1920, and consequently came under the Provincial Minister.[5] With the introduction of the Government of India Act 1935, forests became a transferred subject in all provinces under the control of an elected minister. In the event of the other provinces following this lead, it was believed that the Central Government would have little to do with the forests and their management. However, it was argued by the British officials that the mere fact that forest revenues had become provincial was a strong reason why the Government of India should one way or another control, and so coordinate forest policy, throughout the country. It was believed that only a central authority could settle the broad lines of policy. Forest budgets on which the protection and progress of the forests depended having been transferred to the hands of the Council, it was argued that the Government of India must exert control to prevent reckless or ignorant waste through excessive grazing etc. or trenching upon forest capital by a province which was in need of money. This check they believed could best be maintained through the working plan.[6]

---

with the restrictions imposed by the Forest Department were the real causes behind this problem. However, these causes were not given enough importance in the discussion. *The fodder problem in Bombay Presidency*, Government Press, Bombay, 1914, pp. 1–23.

5 One feature of the Reform Act of 1919 that was constantly criticised by the Congress was, Diarchy under which only certain departments were transferred coming under the administration of Indian Ministers. Other subjects were "reserved" remaining under the control of the Secretary of State for India, and forests was one of these, though in Burma and the Presidency of Bombay, Forestry was soon afterwards moved to the list of transferred subjects.

6 E. P. Stebbing, *Forests of India*, Vol. 3, p. 292.

On the whole the first two decades of the twentieth century witnessed a lot of organisational changes though there were no major changes in the forest policy as such. During World War I when the enormous demands in timber and other forest produce required by the military authorities to carry on the work of war became apparent, a very large amount of additional and urgent work was demanded from the personnel of the department and several provinces were called upon to play an exceedingly important and quite unforeseen part in the war.

A comment made by E. P. Stebbing is very revealing in this matter. He says "It would be difficult to speak too highly of the zeal and energy and patriotism of the officers who successfully coped with the operations required, many of whom had wished to join the forces in the field. The department will forever honour its representatives who away from the limelight and glamour of active service carried out, amidst the solitude and hardships inseparable from such work in great forest tracts, duties in connection with the supplies indispensable to the prosecution of the War."[7]

It was well recognised by the Forest Officers that their department was becoming very much alive to the necessity of commercial forestry.[8] It was believed that a forester who cared to look ahead could see that a real boom in timber trade was anticipated, and hence an old sleepy forest policy had to be done away with. They felt the need to adopt a bolder policy by which forest loans could be raised, employment could be provided to forest engineers, a large forest selling department could be created and an up-to-date method of transportation could be

---

7 E. P. Stebbing, *Forests of India*, Vol. III, p. 531. Another officer very proudly stated that "On the whole, the forest, civil and military officials have worked as one team, though occasionally some red tape wallah has butted in with the letter of the law in his mind, evidently above the spirit of the undertaking, which it need hardly be added is 'Don't waste time but help win the War as soon as possible,'". See "Notes on the Supply of Hay to Our Fighting Forces," *Indian Forester*, April 1918, pp. 168–170.

8 "Indian Forestry: Indian Engineering," *Indian Forester*, September, 1917, p. 425.

formulated. Such and other issues could be handled.[9] The net revenue of the Forest Department had increased twelve-folds in 50 years from 1864 to 1914. It averaged over 1,000,000 pound sterling per annum during the last quinquennial period, i.e. from 1914–15 to 1918–19. This was without including the large sum represented by the value of forest produce given away free or removed by right holders.

Table 1 shows the Financial Results of the Forest Administration in British India from 1864–65 to 1918–19 (in lakhs of rupees averaged per annum).

Table 1

Quinquennial Revenue	Gross	Expenditure	Surplus	% of surplus to Gross Revenue
1864–65 to 1868–69	37.4	23.8	13.6	36.4
1869–70 to 1873–74	56.3	39.3	17.0	30.2
1874–75 to 1878–79	66.6	45.8	20.8	31.2
1879–80 to 1883–84	88.2	56.1	32.1	36.4
1884–85 to 1888–89	116.7	74.3	42.4	36.3
1889–90 to 1893–94	159.5	86.0	73.5	46.1
1894–95 to 1898–99	177.2	98.0	79.2	44.7
1899–1900 to 1903–04	196.6	112.7	83.9	42.7
1904–05 to 1908–09	257.0	141.0	116.0	45.1
1909–10 to 1913–14	296.0	163.7	132.3	44.7
1914–15 to 1918–19	371.3	211.1	160.2	43.1

Source: The Work of the Forest Department in India, prepared under the direction of the Inspector General of Forests to the Government of India, Imperial Forest Research Institute, Dehradun, April 1920.[10]

During the war every effort was made to substitute indigenous timber for foreign supplies, to reduce the demands on shipping to a minimum and to encourage the use of locally grown material. The

9 *Ibid*, pp. 425–426.

10 *The Work of the Forest Department in India*, prepared under the direction of the Inspector General of Forests to the Government of India, Imperial Forest Research Institute, Dehradun, April 1920, pp. 51.

data of the operations of the timber branch from its organisation in April 1917 to the end of October 1918 revealed the total balance of stock in hand in three depots at the end of October 1918 as 42,624 tonnes viz. 22, 693 tonnes (including 3,47,849 bamboos) in Bombay, 13,720 tonnes in Rangoon and 6,211 tonnes in Karachi.[11]

The main type of timber supplied for war purposes was in the form of saw beams, planks and scantlings. Apart from supplying timber and bamboo for structural work like construction of bridges, piers, wharves, buildings, temporary huts and the lines, the Indian Munitions Board provided wood of suitable species for the construction and repair of craft, telegraph poles, bamboo, for river training works, tent and mosquito poles. In addition, the department was engaged in exhaustive search for timber suitable for use in aircraft. Numerous species were tested and possibilities for utilising many of the so-called valueless jungle woods were considered. Enquiries were made by the Research and Experiment Department for woods suitable for rifle stocks, embossing work, brush making, buttons and a particular species found in Canara forests of the Bombay Presidency for aeroplanes.[12] In little over a year (April 1917–October 1918) 228,076 tonnes of timber (excluding railway sleepers) were supplied by the specially created timber branch of the Munitions Board and 50,000 tonnes of fodder grass exported to help military operations in Egypt and Iraq.[13] During this period the teak export trade continued to pay well over one million sterling annually to Britain.[14]

---

11 In order to enable indents to be met without delay – the extraction of timber from the forests being possible as a rule only during certain seasons of the year, and to season the timber before dispatch as far as time allowed – the board established large depots in Mumbai, Rangoon, and Karachi. The figures mentioned here include bamboo but not wooden railway sleepers. E. P. Stebbing, *Forests of India*, Vol. III, p. 532

12 *Ibid*, also see "Indian forestry: Indian Engineering," *Indian Forester*, September 1917, pp. 425–427.

13 Madhav Gadgil and Ramachandra Guha, *This Fissured Land*, OUP, New Delhi, p. 137.

14 *Ibid*, According to the staff of the Indian Engineering Department "A vast number of ships had gone to the bottom of the sea (during the early period

In case of Bombay Presidency, fodder grass operations were carried out by the department with conspicuous success in spite of considerable difficulties involved in their initiation and maintenance. All expenditure was curtailed by the Forest Department and mainly such projects likely to be useful for execution of Munitions Board's indents, were allowed to be undertaken. The Alnawar-Dandeli Railway was constructed as a result of that policy. It was opened for traffic on 1 February 1919. The distance covered was 19.5 miles. Within a year, the traffic on this railway increased to 6,000 tonnes. Traffic in fuel and bamboos sprang up and promised to help the railway line greatly. The opening of the railway made the Dandeli Timber Depot a very important station. The success of the line depended on the ability to export timber and other goods. Shortage of wagons often interfered with the profits of the railway. The Alnawar line was an important means of transport in one of the most densely wooded districts of India.[15]

Since trade in timber was the main commercial activity of the forests, saw mills were almost indispensable. The main product of the mills was railway sleepers for which the demand was constant. In 1900, a saw mill in Ittanbail, in Canara turned out 52,173 sleepers valued at Rs. 85,108. The mill showed a surplus of Rs. 46,042 in 1902. The Forest Report of 1907–08 stated that the saw mills were one of the most paying concerns of the Forest Department. The production of saw mills was

---

of War) and it was likely that many more would follow the same fate. For this purpose, they had embarked on a policy of frantic ship building to replace losses. There must consequently be for many years a sustained demand for ship building timber of every description, of teak in particular, which has become a favourite timber in European ship building trade. Had the Forest Department of the Government of India been already organised so as to meet the coming demand there would have been large profits in store for it; but it is not too late yet." See "Indian Forestry", *Indian Forester*, September 1917.

15 See R. D. Choksey, *Economic Life in Bombay Konkan 1818–1939*, Bombay, 1957, pp. 142–143; Also see *Audit Report of the Accounts of the Bombay, Government Saw Mills at Dandeli, 1942–43*, Government Press, Bombay, 1945, p. 47.

teak sleepers, planks, slabs, scantlings and battens. There was further expansion in the saw mill business due to increasing war demands. During the war large quantities of saw mill output, besides being sent to Mumbai, sent to distant markets such as Egypt and Mesopotamia.[16] However, some of the enthusiastic staff of the British government were not satisfied with this expansion. They stated "Looked at it in any way the question may be taken as settled, that there must soon be a greatly increased demand for timber in all parts of the world – for restorations, for a mercantile marine, for railways – and it will be a piece of folly for India to sit with folded hands while the demand is being met by others." They further stated "Outside India are we not making ourselves responsible for building in the near future railways in Mesopotamia, in Persia, even in Egypt. The sleepers and other timber for such railways should naturally come from India."[17]

The turnout of timber and fuel in Bombay Presidency increased during the War years in Table 2.

Table 2

Year	Total Turnout of timber and fuel in cubic feet	Average Turnout per square mile in cubic feet
1913–14	46077346	3764
1914–15	41539419	3341
1915–16	50184778	4021
1916–17	–	–
1918–19	56888000	4520

Source: Quinquennial Review of Forest Administration in British India for Period 1913–14, 1914–15, 1915–16, 1918–19.

This table shows almost 23 per cent increase in the total turnout of timber and fuel during the 4 years of war. The total turnout of timber and fuel shown in the table represents the turnout of timber and fuel from the reserve forests, protected forests and unclassed forests.

---

16 R. D. Choksey, *Economic Life in Bombay Konkan 1818–1939*, Bombay, 1957, p. 144.

17 "Indian Forestry", *Indian Forester*, September 1917, pp. 423–424.

In 1916–17 extraordinary fellings were made in most of the suitable areas of Canara West and Belgaum divisions for supplying the various types of wood to the Munitions Department; for example junglewood poles 21,131 (No.), logs 9,593 cubic feet, mango planks 4,620 running feet and bamboo of sorts 2,01,150 (No.). In 1917–18 the following produce was supplied to the Munitions Department: Teak logs 1,28,534 cubic feet, battens 1,00,000 running feet, junglewood logs 83,187 cubic feet, poles 74,199 (No.), scattlings 471 cubic feet, fuel 8,493 tonnes of 100 cubic feet each, bamboo 1,374,700 (No.) of different sorts, Tarward bark 76 tonnes from Dharwad and Bijapur, hirda 75 khandis from Canara (1 khandi is equal to 560 pounds).

During 1918–19, the following material was supplied to Munitions: Teak logs 19,853 cubic feet, poles 888 (No.), scantlings 557 cubic feet, battens 138,888 running feet, junglewood logs 76,825 cubic feet, poles 100,730 (No.), scantlings 3,152 cubic feet, battens 363 cubic feet, fuel 2,94,609 cubic feet, battens 363 cubic feet, fuel 2,94,609 cubic feet, bamboo 53,243 (No.), tent pegs 5,220 (No.), Tarwad bark 14 tonnes and 13 mounds.[18]

By 1923, the area under the control of Forest Department in Bombay Presidency was 12,504 square miles and 12,98,359 square miles were brought under the survey.[19]

Strangely enough, the meeting of war demands by excessive felling in the forests was considered very patriotic and heroic, whereas people enjoying forest produce to meet their subsistence needs was a crime! In fact with the cessation of the war when the revenue from timber fell, the Forest Department fought to create interest in Indian timber in England and elsewhere.[20] Adding to that the labour situation turned bad due to the influenza epidemic. Wholesale buyers were overstocked with timber bought at higher prices. This coupled with shrunken credit and import of cheap timber from Siam, Japan and America kept the Indian market stagnant. At this juncture the Forest

---

18 E. P. Stebbing, *The Forests of India*, Vol. III, London, 1926, p. 554.

19 *Ibid*.

20 R. D. Choksey, *Economic Life in the Bombay Konkan 1818–1939*, Bombay, 1957, p. 151.

Department tried to create interest in the Indian timber in England and elsewhere. Nevertheless, due to coal strike and other reasons, London remained a lukewarm market, though in 1920–21, 650 tonnes of teak was sent to England. Labour and transport difficulties along with the Great Depression in 1929 did not allow the timber trade to prosper to the extent the contemporary Forest Officers aspired for. However, the efforts to improve profits through timber exploitation went on incessantly. For example, at the beginning of the twentieth century meager sums were spent on road construction in the forests of the Konkan. In Thana Rs. 626 were spent mostly on road repairs, in Canara Rs. 1,047 and in Kolaba Rs. 759. Five years later, as indicated in the Forest Report of 1905–06 a total of Rs. 8,350 for communications in forests. In 1910–11 a scheme of roads for the Northern Circle, which included the Thana forests, was worked out at a cost of Rs. 75,000. In Canara, inclusive of a new road, the cost for the year 1910–11 was Rs. 28,854. Two years after the First World War, according to the Forest Report of 1920–21, a complete road programme for Pint, Thana and the Dangs was drawn up and awaited an adequate engineering staff and funds to carry it out for the next few years. In 1925, several roads were projected with the intention of throwing open inaccessible parts of Pint, Thana and the Dangs. Work continued on some of the projected roads in Canara. By 1930, the total expenditure on roads by the Forest Department amounted to Rs. 2,69,834.[21] Copleston, the Conservator of Forests, wrote in 1924 that the extension of road communications was and is bound to be the main line by which Forest Department can advance.[22]

Another important change brought about in the Forest Department during this period was that in 1927 the Forest Act XVI was passed. Till then the Indian Forest Act of 1878 was the model for corresponding Forest Acts of the majority of the Commonwealth countries. This Act was revised in 1927 as Forest Act XVI.[23] The changes made by this Act were small.

---

21 *Ibid*.

22 *Ibid*, pp. 141–142.

23 The Law Committee gave the following statement of objects and the reasons of the Act. "The general law relating to forests in British India is contained

In Section 30 (b) permission to enclose portions of Protected Forests and suspend rights therein for a maximum period of 30 years compared to the 21 years fixed in the 1878 Act, Section 79 clarified and made more extensive the duties of the public to prevent and extinguish forest fires and help forest or police officers in preventing forest offences and the exercise of control over private forests where necessary.

However, no action was taken until 1939 when a new bill was drafted. Its enactment was delayed by the war but subsequently many of the provinces passed Private Forests Control Acts, for example Bengal, Bihar, the United Provinces (1947), Madras (1946), while Mumbai passed a more comprehensive Bill in 1942 (Act XXVIII). Their underlying principle was to manage forests according to an approved working plan with the alternative of having the forests being taken over by the State.[24]

Enactments and regulations or rules were made under the Indian Forest Act by several provinces and were referred to under their respective chapters. These enactments related to the private forests (for example in Bengal, Bihar and the United Provinces) as the provisions of the Indian Forest Act, were found insufficient in preventing the rapid decline of the forests into wastelands with its consequent dangers.

The most important rules and regulations made under the Act by the provinces, were perhaps those controlling the transit of forest produce. Burma and Madras Presidencies were not included in the

---

in the Indian Forest Act, 1878 and its amending Acts. The present Bill brings the law together within the scope of one enactment. The Bill is a straight forward consolidating Bill but the original Act having been passed before the general Clauses Act of 1897 (X of 1897), it has been possible to shorten the language of the Bill by taking advantage of the Act. The ambiguous language of the second paragraph of Section 42 of Act VII of 1878 has been altered in clause 42 (2) so as to bring it into conformity with what appears to have been original intention of the Law", *Gazette of India*, 1926, Part V, p. 165.

24 C. B. Upadhyay, *Forest Laws*, Hind Publishing House, Allahabad, 1984, pp. 4–7.

Indian Forest Act and had their own Acts.[25] For example, the Punjab Act II of 1900 was passed as the Punjab Land Preservation (Chos) Act. This Act was enacted to preserve and protect territories adjacent to the Shivalik Hills affected by the debasement of forests or by the action of streams and torrents (Chos). This Act therefore affected the work of the Forest Department considerably.

However, another important point of general concern during this period was the maintenance and improvement of fuel and small timber supplies for local consumption. With the struggle for Independence assuming a more serious form it had become necessary for the government, by this period, to give consideration to people's needs and complaints. The government could no longer ignore the popular public opinion. For example in case of Bombay, especially after the Forest Department had become a transferred subject in that Presidency, the government had to take serious cognizance of the peoples' complaints. The Bombay Government appointed a Forest Grievance Committee in accordance with a resolution passed in the Legislative Council on 30 July 1924 under G R No L C 1374, dated 26 January 1925. The Committee's report was released in 1926 and the government accepted 19 recommendations after modifications, while 25 were rejected as unreasonable. The majority of the concessions allowed concerned grazing, collection of grass, fuel and minor produce.[26]

The peculiar feature of this report of the Committee was that it was not unanimous owing to the wide difference between the views of the official and non-official members. On many very important points it was found impossible to write a unanimous report and hence the Committee finally decided to submit two separate reports, one representing the views of the official members and the other those of the non-officials. Edie, the Chief Conservator of Forests and Conservators of the 3 Circles viz., North, Central and South were the

---

[25] Punjab Act II of 1900 was passed as the Punjab Land Preservation (Chos) Act.

[26] *Forest Grievance Enquire Committee Report*, Government Press, Bombay, 1927, pp. 1–76.

official members. The non-official members were the members of the Legislative Council. The Committee was empowered to visit important forests situated in the Presidency proper, and to enquire into the grievances of the *rayots* with regard to forest matters and to submit suggestions for the consideration of the Government.

An example of the wide difference between the perceptions of the official and non-official members was the issue of tackling forest fires. The *rayot* thought the communal punishment imposed by the Forest Department in case of failure in extinguishing forest fires or preventing them was unjust. Over this the official members commented that if people insist on setting fire to the forests, public opinion must be aroused against the system of firing and that can be done only by the combined system of punishment and rewards. Whereas the non-official members found it very unjust and rightly resented by the people. They commented that power given to the Forest Officers was more often abused than properly exercised. They acknowledged the fact that the people had come to look upon the vicinity of forests to their villages as a curse rather than a blessing. These non-official members remarked that the local officers ought to be more sympathetic.[27]

The comment made by E. P. Stebbing over the report of this Committee is typical of a British officer, who assumed to possess a monopoly over the issue of knowledge regarding forest conservation. According to Stebbing and many of the Forest Officers, "the concessions demanded by the villagers could not have done any damage individually but in the aggregate they would have accelerated the degradation of the poorer forests where restrictions were most needed for the maintenance of productivity in the villagers' interests. Some of the concessions concerning cultivation definitely constituted a loss of forest area".[28]

That the widening of the gap between the forest management and the subsistence practices of the villagers had reached its climax

---

[27] *Ibid*, pp. 1, 16, 17, 31 and 32.

[28] E. P. Stebbing, *Forests of India*, Vol. IV, John Lane, London, 1921, p. 255.

point is obvious from this report.[29] One of the suggestions of the Grievance Committee, which was accepted, was to create Forest Advisory Committees in 8 districts to promote regular discussion of questions concerning the local administration of the Forest Department, and remedying any legitimate grievances of the public. These Committees composed of the Collectors (as Chairmen), the Divisional Forest Officer, one or two officers of the Agricultural or Cooperative Department and a number of non-officials, including all members of the Legislature residing in the district.

Not satisfied with the concessions granted by the Grievances Committee of 1925, another Forest Grievances Enquiry Committee was appointed in 1935. Its recommendations led to a further series of concessions.[30] The change in the political atmosphere in the country definitely encouraged people to come forward with their grievances, and concessions continued to be granted from time to time, despite unwillingness and discontent among the Forest Officers.

In 1928, the Royal Commission on Agriculture in India made inquiries into the uses of forest lands which were under the management of the Forest Department for agricultural purposes such as providing fodder for livestock, fuel and timber for the rural population and protection of soils liable to erosion. In its report in 1928, the Commission, taking into consideration the declared policy (1894) of the Government of India, made certain recommendations which were quite sympathetic towards the rural folk as compared to the attitudes of the government officers seen in the past. However, the basic assumption was the same, that the cultivators and other rural folk had wasteful habits as far as use of natural resources was concerned and they were prone to take a short-sighted view of even their interests.

---

[29] Richard Tucker, "Forest Management and Imperial Politics", Thana district, Bombay, 1823–1887, *The Indian Economic and Social History Review*, Vol. 16, No. 3, pp. 298–299.

[30] G.R.R.D No. 3449/33 of 24/238, Forest and Agriculture Department, Maharashtra State Archives, Bombay.

The Commission indicated complete agreement with the general forest policy laid down by Dr. Voelcker on the point that "in regard to both fuel and fodder reserves and grazing areas, pure and simple, especially such as lay in the midst-off cultivated tracts, they asked local governments to consider whether it was necessary to class these, or if they were already so classed, to retain them as forest areas; and if this question was decided in the affirmative, whether it would not be better to constitute them protected rather than reserved forest."[31] Here, the general relaxation of restrictions was recommended by the Royal Commission with the intention to promote agriculture to increase revenue earnings.

The comments made by the Commission regarding the acquisition of forest land by the Forest Department in its early phase of the administration are very revealing. The Commission stated "40 or 50 years ago, when there were Forest Departments, but no agricultural departments and when the problems arising out of the pressure of population on the soil and the necessity of the cattle improvement had not assumed the importance they now possess, it was natural that the Forest Departments should be regarded as the only department capable of turning to advantageous use large areas of State land, which were lying neglected and which were not required for cultivation. The result was that the presence of tree or scrub jungle was considered sufficient justification for notifying as forest land, which was in reality more suited for grazing or for cultivation than for the growth of trees..."[32] This comment unintentionally admits the over enthusiasm of the Forest Department in acquiring as much land as possible, in order to earn more and more revenue. However, the Commission seems to be oblivious of the misery caused by the over enthusiastic Forest Department to the common people.

---

31 The Royal Commission decried the protected forest as a forest where is nothing is an offense that is not prohibited. Dr. Voelcker, *Report of the Royal Commission on Agriculture in India*, Mumbai, 1928, Reprint Agricole Publishing Company, New Delhi, 1982, pp. 258–260.

32 *Ibid*, p. 269.

In the Commission's perception, the most promising method of establishing village forests was to handover to village management or village panchayats certain wooded areas, which were under the management of the Forest Department. Though in Mumbai the Commission found divided opinion on the question of feasibility of this kind of popular management of forests, the Commission stated that the management by the people of the forests close to their villages possessed so many desirable features, that every effort should be made to ensure its success.

Management of forests by village panchayats was introduced in some districts of Bombay Presidency. However, only a few panchayats survived their term, for example 4 in North Canara and 13 in West Canara. Barring Assam and Bombay, in other Presidencies, for example in Madras, the panchayats gave satisfactory results.

In spite of the several recommendations made by the Royal Commission, the issues related to fuel and timber supply, forest grazing and fire protection continued to bother the forest officials. They became critical of the tendency of Congress ministry to grant several concessions to the *rayots*. In the year 1940–41, the Provincial Fodder and Grazing Committee met on 6 June 1940 and discussed issues like survey of grasslands and new grazing rules. As a result of the Committee's recommendations two schemes were approved. One was grass drying scheme and the other was reclamation of marsh lands. Also during the year under report, each district of the province established a District Fodder and Grazing Committee except Sholapur district.[33]

However, one of the major landmarks in the history of the Forest Department in British India was World War II, during which exploitation of forests took place at an unprecedented scale. The Forest Officers who constantly criticised the village population for having destroyed the Indian forests, competed with each other to praise the

---

[33] The Committee considered two experiments for the improvement of the grasslands to get leading agriculturists to experiment a substitute for grass. To let out government land purely for growing fodder crop at concessional rates for some period.

contribution of the British Indian Forest Department in World War II.

In contrast to this criticism of people and the Congress Ministry for giving the people various much needed concessions, E. P. Stebbing commented "The Forest Department was staffed by men who were not consciously in the public eye either in peace or war. Yet its contribution to victory was by no means insignificant. It was called upon to carry out a three-fold task; firstly to provide weighty material support to the armed forces in the field; secondly to administer to the greatly increased needs of industry and thirdly, to carry out the day to day administration of public affairs."[34] Stebbing gladly accepted the compromises made in case of conservation of forests due to the gigantic war supplies of forest produce. He said "It was inevitable that the insistent demands of War called for measures which at times threatened to sweep aside the basic tenets of conservancy and sustained yield in forestry. During those years, millions of tonnes of timber crops were selected, felled and converted into a great variety of munition. For the entire range of forest product India could depend solely upon her own resources. If the Indian Civil Service was the 'steel frame' of British Administration, the 'wooden frame' of the Forest Service proved equally tough and resilient."[34]

During this period, almost all sections of the Forest Research Institute were dedicated solely to the requirements of war and all inessential work was suspended. Compliance with orders for all classes of material entailed both more intensive exploitation of readily available resources and an extension of fellings into previously inaccessible tracts.

From the west coast tracts of Bombay, Madras, Coorg and Cochin the vast demand for sleepers and other timber was met not only from deciduous forests but also largely from evergreen rain forests by exploiting timber never before utilised. Timber production was about 2,42,000 tonnes in 1940–1941.

---

[34] E. P. Stebbing, *Forests of India*, Vol. IV, John Lane, London, 1921, pp. 144–147.

Table 3 shows the war supplies made from Bombay Presidency.

Table 3

Description	**Material Supplied**	
	Quantity	**Value (Rs.)**
	Northern Circle	
Ballies 6"	5000	100700
Ballies 4"	145000	
	Central Circle	
Ballies 4"	41780	31630
	Southern Circle	
Fitted bamboos for tent poles	3000	7131
Bamboos for war reserve	8245	4889
Sheebs for aircraft making in England	5000	426
Teak ballies	11708	12879
Rosewood logs	80 tons	8768
Bijasal (honne logs)	20 tons	
Dhamin logs	67 tons	5225
Dindal logs	50 tons	
Nailable timber in hutting sizes	100 tons	75000
Teak logs squared	400 tons	66000
Teak logs round	160 tons	
Teak spars	100 tons	
Teak squares	12 tons	2596
Teak spars	55 tons	
Teak squares	12 tons	
Teak transmission poles	345 tons	9716
Teak sleepers of special sizes	55 tons	1375
Teak squares	100 tons	12500
Teak squares and sawn material	4636 Cft.	22275
Total		361110

Source: *Forest Administration Report of The Province of Bombay, 1941–1942*, Government Press, Bombay, 1942, pp. 31–32.

A comparative statement of major and minor forest produce extracted departmentally in the Bombay Presidency during the 2 years of 1939–40 and 1940–41 also shows an increase in timber from 936 thousand cubic feet in 1939–40 to 1329 thousand cubic feet in 1940–41, due to extensive departmental operations undertaken for supply of ballies (poles) from the Northern and Central Circles and of logs and sawn material from the Southern Circle for war purposes, and also for supply of a larger quantity of sleeper pieces required by the mills of Dandeli, to meet the increased demand for M. G. Teak sleepers of the M. and S.M. Railway in the following year. The supply of bamboos to the Defence Department was responsible for the increase under this category. In the Bombay Presidency it was Rs. 43 in 1939–40 which increased to Rs. 2724 in 1940–41.

Compared with the figures of the preceding year the revenue, expenditure and surplus increased during the year 1940–41 by 593.1, 172.6 and 420.4 thousands of rupees, respectively. One of the chief causes of this was supply of timber and bamboos for war purposes. During this period, the expenditure under conservancy and works increased by 149.2 thousands of rupees. This was due to the extensive departmental operations undertaken mainly for the supply of timber and bamboos for war purposes and for some other purposes like sawing up a large number of sleepers for supply to the M. and S. M. Railway in 1941–42. During this period, timber and fuel were extracted through the agency of purchasers by the sale of standing coupes in the Northern and Central Circles. Fuel for the Government was extracted departmentaly in both Circles. During the year, the Forest Engineer also had to extend his activities of war supply work. For an order of 3,000 fitted tent poles and treated bomboos he had to design a special sheet steel tank for treatment of bamboos with oil and creosote and presses for strengthening them.[35]

Among the usual commercial activities the supply of sleepers to the Railways continued to receive careful attention obviously because, a strong and efficient railway network was required to fulfill the needs

---

[35] *Ibid*, 34.

of the war. E. P. Stebbing has given due credit to the Railways for responding magnificently to the needs of the hour.

During the war, felling and sawing was pushed into the remotest forests of the Himalayas and into the densest jungles of the Western Ghats. Working Plan prescriptions in reserved forests were considerably upset during the war. In Bombay, the yield prescriptions departed from the margin to the extent of 400 percent.[36] It was not that only direct supplies for the war caused all the devastation of forests. There were some indirect causes too. The shortage of commodities created due to the war, in turn exacerbated the process of destruction of forests. For example, with coal already unobtainable and petrol severely rationed a fuel famine ensued during this period in many parts of India. Lacking coal, many industries were compelled to resort to firewood for power. Motor transport turned increasingly to the use of producer gas, particularly in the Madras Presidency. This development necessitated the conversion of large quantities of wood to charcoal and resulted in excessive cutting of forests in Madras. By 1944 intense demand for fuel had arisen in urban areas of all provinces. Consequently, fuel rationing was enforced in all heavily populated regions.

Firewood meant for civilian needs was delivered to the civil district authorities for retail distribution and was sold at authorised price. However, government reserves were able to provide only a portion of the total supply of firewood. By 1944 local and private sources of supplies had already been so heavily exploited that they were almost exhausted. Consequently, town people became more dependent on supplies from the reserves.

Industrial timber too was not immune to this. With the ever growing and insatiable demands by the Supply Department, fulfillment of the demands of civilians and trade became impossible. The quantity that arrived for trading purposes was a very insignificant fraction of the total demand. In effect this too contributed to the grave destruction. The danger of war had receded by May 1945. But the supply of

---

36 Madhav Gadgil and Ramachandra Guha, *This Fissured Land*, OUP, New Delhi, p. 134.

materials to the Defence Department did not relax until the end of 1945.

Such was the havoc played with India's forests during the war and yet according to Stebbing, "In the final outcome the forest legacy was preserved essentially intact, albeit reduced, in striking contrast to the widespread devastation of forest lands which lay outside the jurisdiction of the Department."[37]

Here an essential question comes to the mind as to what justified such exploitation of these forests which, according to the Forest Officers of that time, were already devastated and needed immediate care and protection from the villagers. All along the history of British forestry the most vocal complaint found against villagers was that they had utilised the forests to such an extent that they were seriously devastated. Then what moral justification could be offered to the exploitation of already destroyed forests of a colony.

---

[37] E. P. Stebbing, *The Forests of India*, Vol. IV, p. 147.

7

People's Resistance

> "...The new forest rules have fallen like a fatal blight – like a thunderbolt from a clear sky. By one direful stroke of the pen 'the man of the jungles' finds himself at once a proscribed outcast in his own wilds. His hills and jungle fastnesses are suddenly proclaimed to be State forests. Every vegetable and mineral substance therein is declared to be 'forest produce.' All forest produce is declared to belong to the Crown. And no one is allowed to move any forest produce whatever without the formal permission of the 'jungle-wallah sahib,' the new forest King..."
>
> A letter to the Editor
> "Forest Conservancy in Bombay", *Indian Forester*,
> September, 1886, p. 303.

The British Forest Department and its forest policy had an everlasting and far-reaching impact on the Indian society. A society, which was primarily agrarian in nature, had a very close dependence on forests. It was not only the forest dwelling tribal communities which were dependent on forests but the entire rural population, agriculture being their main source of income, was dependent on forests in some way or the other. With their sustainable style of living, in the normal conditions poor peasants were dependent on forests for their supplementary needs. Artisans depended on them for their raw material. In case of natural calamities too, forest was the most important alternative to these communities on which they could rely for their survival. Also they could migrate from one place to another in case of famine or drought or any other disaster.[1] Land without cultivation did not have any commercial value. However, with the establishment of the Forest Department, the poor people lost their supplementary source of livelihood. Keeping cattle which was an essential component in the life of the cultivator became difficult due to the restrictions imposed by the Forest Department on grazing. Artisans lost their source of raw material from the forests due to forest restrictions.

Adding to this, the machine-made goods brought from England left the artisans redundant. They had nowhere else to go but to agriculture. Consequently, pressure on land for cultivation increased tremendously. The Census Reports of 1891 and 1901 for example indicate this pressure on land (see Table 1).

Table 1

Occupation	Population			
	1891	1901	1891	1901
Agriculture	10951000	10991000	60.7	61.9
Industrial	3730000	3149000	20.6	17.7
Commercial	607000	490000	3.3	2.8
Professional	373000	277000	2.0	1.6
Non-agricultural labour	630000	1078000	3.5	6.1
Others	1744000	1754000	9.9	9.9

Source: *Memorandum on the Material Condition of the People of the Bombay Presidency 1892–1902*, Bombay, 1902.[2]

1 This dependence of the rural population on forests was acknowledged even by some of the British officers for example, "Forestry has been termed the handmade of agriculture and nowhere does this apply with greater force than in India, which is essentially an agricultural country. Apart from their ordinary domestic and agricultural requirements for timber, fuel, thatching and fodder grass, grazing and numerous minor products, in many parts of the country the people are largely dependent on the forests for their very existence in times of famines." *The Work of the Forest Department in India*, prepared under the direction of the Inspector General of Forests to the Government of India, Dehradun, April, 1920.

2 *Census Reports of the Bombay Presidency*, 1891 and 1901.

Table 1 clearly shows decrease in commercial population by 117,000. Agricultural occupation shows a remarkable increase. Depression in other means of livelihood, must be mainly responsible for the increase in population supported by agriculture. Also, there is a decrease in the number of professionals which include the artisans, traditional religious mendicants, etc. This increase in pressure on land for cultivation, meant worsening of the prevailing circumstances for the cultivators as due to their indebtedness, large part of cultivable land was getting transferred from the hands of the cultivators to the non-cultivators.[3] The British Government recognised the right to transfer land as a part of the right to property. Land records prepared and maintained under the British Government facilitated such transfers. The Bombay Land Revenue Code (Act 5 of 1879) explicitly recognised the right to sell, mortgage, bequeath or otherwise to transfer land (Section 73)[4]. For example, in Thana almost all the land became the property of non-cultivators.[5]

M. G. Ranade, the prominent nationalist leader of his times, described the wretched conditions of the agriculturist in the nineteenth century, after a careful survey based on the 1872 Census in the following manner

---

3 This misery brought to the rural population due to the forest restriction has been described very well in a Marathi book *Gao-gada.* Its author was a government employee in the pre-independence period. This profession required him a lot of touring in the country side of Bombay Presidency. He had received training under Mr. Enthoven, Superintendent of Anthropological Survey of Bombay Presidency, see T. N. Tare *Gao-gada* (Marathi), Third Print, Bombay, 1959, pp. 118–138.

4 S. D. Kulkarni,"Alienation and Restoration of Adivasis' Lands in Maharashtra" *Economies of The Tribes And Their Transformation,* (ed.), K.S. Singh, New Delhi, 1982, pp. 420.

5 J. A.Voelcker, *Report on the Improvement of Indian Agriculture,* 1893, reprint 1986, pp. 292. Also see Symington D. Report on the Aboriginal Hill Tribes of the Partially Excluded Areas in the Province of Bombay, Bombay, 1939. Also see, for the comments made by T.D. Mackenzie, the acting Collector of Thana, Revenue Department, Vol. 33, 1884, Maharashtra State Archives, Bombay, p. 17.

"We have thus a poor soil (he was referring particularly to the Deccan) afflicted with scanty and irregular rains, inhabited by a sparse population, for the most part agricultural and uneducated, with no openings for labour save an exhausted soil, with average holdings of less than 10 acres per each head of family, burdened with a payment of State demand which represents one-third of their net gains, forced to be content with a hand-to-mouth subsistence represented by a maximum of Rs. 60 a family, and involved in heavy bonds of debt under circumstances beyond their control to a numerous mass of small creditors, largely foreign in their domicile, and not prepared to take up cultivation of land on their own account. These are the salient features of the social and economic condition of the people which, is necessary to bear in mind, while discussing the question of the best method of relief."[6]

The description given above by M. G. Ranade, shows briefly the wretched conditions in which the Indian agriculturist had fallen into during the nineteenth century. Naturally, it had given rise to a lot of discontent in the society. Deprivation of age-old rights over forest lands had demoralised the people. The forests in which they had been dwelling since time immemorial were suddenly alienated from them. Now if they were found dwelling in these forests, it was considered an act of offense.[7] The acts which were their rights earlier had become offenses now, for which heavy fines, or some other punishments like imprisonment were imposed upon them.[8]

---

6 M. G. Ranade, "Land Law Reforms and Agricultural Banks", *Builders of Modern India,* (ed.), K. T. Paul and N. Macnicol, Bombay, 1926, p. 35.

7 A tribe looks upon the forests as it's natural environment and starts depending on it. This dependency syndrome engenders growth of certain features which characterise the tribal society and it's economic system, see V. C. Childyal, "Tribe and Forest", *Economies of The Tribes And Their Transformation,* (ed.), K.S. Singh, New Delhi, 1982.

8 As one of the letters to the editor of *Indian Forester* puts it, "Does a Varli scratch clean half an acre of slope and cover it with a layer of bushes and scrub all ready to burn – down comes the forest guard and arrests him for committing waste! Does he lop a kheir or an ain tree, or any of the hundred and kinds specially reserved – away with him to the magistrate for injuring

Commercialisation of forests had led the Forest Officers, timber merchants and contractors deep into the forests.[9] The traditional forest dwellers could see these timber merchants and contractors felling trees and taking away the timber to town markets.[10] However, the same action was not permitted to these forest dwellers. It was thought that protection to forests and maximisation of revenue would be difficult if too many privileges were granted to tribals in the forests. So much so that Shuttleworth, one of the senior Forest Officers, commented about the *Bhil* of Khandesh that, "... as long as the present system exists of permitting extensive private rights and privileges – it is opposed to the dictates of common sense to expect an efficient protection of the forests on a proper realisation of revenue." For this comment the government had to make him realise that, "Mr Shuttleworth ought to know that it is of much greater importance to keep the *Bhils* quiet, to allow them an honest mode of livelihood suitable to their wild habits, than to take away their means of support and render them outlaw on society."[11] The surplus of the revenue of the Forest Department increased from Rs. 1,99,317 in 1873–74 to net revenue of Rs. 31,70,000 in 1922–23, which shows almost fifteen times increase over the period

Crown property! Does he cut a few reeds for his hut or bamboos for his cattle shed – he is a thief (God save the Monk), he has stolen public property!" See the letter "Forest Conservancy in Bombay", *Indian Forester*, September, 1886, p. 303.

9 Following was one among the thoughts expressed by some of the officers regarding making the inaccessible forests, accessible, "...It is often easy to see on the spot that a few hundred rupees worth of blasting and levelling would make a whole series of inaccessible coupes saleable, but it is not easy to put the case convincingly..." This particular thought was expressed by the Acting Collector of Thana, who was willing to extend his support on general as well as departmental grounds to the Divisional officer for this purpose. See *Forest Administration Report of the Bombay Presidency*, 1901–1902, p. 23.

10 Similar views have been expressed by Haimendorf. For details see, Christoph von Furer-Haimendorf, *Tribes of India*, Oxford University Press, Delhi, 1985, pp. 81–83.

11 Communication by a Forest Officer, "The Budget Estimate of the Bombay Presidency for 1876–77", *Indian Forester*, April 1876, pp. 90–91.

of 50 years.[12] Table 2 shows in rupees the increase in forest returns in Konkan region alone from 1900–1901 to 1933–36 (see Table 2).

Table 2

Years	Revenue	Expenditure	Surplus
1900–1901	1226036	666977	560059
1905–1906	1772343	762614	1010759
1910–1911	2050594	918390	1130204
1915–1916	2635340	1078458	1556863
1920–1921	3261154	1730764	1630380
1925–1926	3750404	1871917	1878487
1935–1936	3278828	1374255	1904573

Source: R. D. Choksey, "Forests", *Economic Life in Bombay Konkan, 1818–1939*, Mumbai, 1957, p. 152.

Such a situation led to a general feeling among the people that unless money was paid nothing could be obtained from the forests.[13] As the Marathi daily *Kesari* quoted there were around 20 lakh tribals in the Bombay Presidency in the 1880s.[14] Naturally these tribals were closely linked with these forests and hence they were the worst affected lot among the total population. As the same Marathi daily puts it, "It is not even 6 years since the Act has been passed. But people have been harassed so much by the Act that it is not difficult to believe if there is a news of unrest among the people anywhere. Prevalence of such circumstances in the State is not healthy either for the rulers or for the subjects..."[15] There were ample number of such unrests all over the Presidency during the nineteenth and the first half of the twentieth century. Early examples of this in case of Bombay Presidency were the revolt of the *Varli* tribals in early 1850s, revolt of *Mahadeo Kolis* and

12 See *Administrative Report of the Bombay Presidency*, 1873–74 and 1922–23.

13 *Kesari*, 26 May 1885.

14 This figure was based on the Census Report. *Kesari* (Marathi daily) 2 June 1885. This Marathi daily was started by the famous militant leader, Lokmanya Tilak.

15 *Kesari*, 21 April 1885.

Bhils in the first half of the nineteenth century. In the words of K. S. Singh, "... The first phase (1795–1860) of the tribal movements coincided with the rise, expansion and the establishment of the British Empire. It saw the rise of what we may call the primary resistance movements. Resistance is inherent in all movements but during this phase it was spontaneous, elemental and widespread involving not only one tribe but many... It was a resistance to the new system and to the new classes of the people who were inducted by it, namely, the system of local administration and taxation, evangelisation and humanitarian measures, the landlords, moneylenders and government officials, all of whom were to be thrown out in a violent upsurge..."[16]

In case of *Varli* revolt, hundreds of *Varli* tribals marched to the government offices to protest against the infringement of private commercial rights on forest produce. The protest got no gains for the tribals. It was brushed aside by officials with the accusation that timber merchants were instrumental in organising this protest. In fact this incident was considered as an example of the danger posed by urban interests organising peasant protests.[17]

Mahadev Kolis and the *Bhils* were the two other tribal communities of the Western Ghats of the Bombay Presidency who revolted against the new rulers. Out of their prominent revolts which took place during the nineteenth century the first revolt was of Junnar in Poona district in 1830. It was led by Bhao Khare, Chimnaji Darbare and Jadhav. Fifty-four *Mahadev Kolis* and their two prominent leaders died in this revolt. Ramachandra Ganesh Gore and many others were hanged to death.[18]

Raghoji Bhangare was a famous hero of the *Mahadev Kolis,* who fought for the people's age old rights over their motherland (which were predominantly inclusive of the rights over forests as tribals belonged to the forests). He gave a brave fight to the Company

---

[16] K. S. Singh, "Colonial Transformation of Tribal Society in Middle India", *Economic And Political Weekly,* 29 July 1978, pp. 12–29.

[17] *Bombay Forest Commission Report*, Vol. I, 1887, pp. 20–27.

[18] Dr. Govind Gare, *Sahyadriteel Adivasi Mahadev Koli* (Marathi), Sahadhyayan Publishers, 1974, pp. 54–60. For further details about Mahadeo Kolis, see A. Mackintosh, "An Account of the Tribe of Mahadeo Kolis", *Transactions of the Bombay Geographical Society*, Vol. 1, Bombay, 1836–38, pp. 234–235.

government. However, he was arrested and later hanged to death by the police in 1848.[19] However, the Thana District Gazetteer describes him as a criminal who plundered merchants and British officers, who had committed thefts and carried on theft of opium. The Gazetteer describes that "His terror was so great that the people of Bhiwandi town shut themselves in their houses."[20]

The revolt led by Bhagoji Naik in 1818, witnessed death of 10 British soldiers and many were injured. Bhagoji Naik with the support of *Mahadev Kolis* and *Bhils* continued to fight against the British until 1859. He died along with his 45 other soldiers in a brave fight against the British near Sinnar. The tribals (*Mahadev Kolis* and *Bhils*) were proud of Bhagoji's achievements against the British. They continued to support him from 1818 to 1859, that is until his death. The British Government appointed special officer Major Franksacher to win him over.[21] However, the revolts which took place during later half of the nineteenth century were directed mainly against the *savkars*, that is, moneylenders who were the creation or product of British rule, their administrative policies and land settlement patterns. Taking advantage of the new policies, *savkars* were flocking from cities to rural areas. Owing to the new revenue policies, land settlement patterns, restrictions on the use of forests and loss of forest lands, tribals were increasingly falling into the moneylender's debt traps.[22] They had lost

---

19 David Hardiman, "Community Patriarchy, Honour: Raghu Bhangre's Revolt", *The Journal of Peasant Studies*, Vol. 23, No. 1, October 1995, pp. 90–91.

20 *Gazetteer of the Bombay Presidency*, Vol. 13, Thana, Bombay Government Central Press, 1882, pp. 636–637.

21 Dr. Govind Gare, *Sahyadriteel Adivasi Mahadev Koli* (Marathi), Sahadhyayan Publishers, 1974, pp. 63–65.

22 For more details see, Ravindra Kumar, *Western India in the Nineteenth Century: A Study in the Social History of Maharashtra*, London, Routledge and Kegan Paul, 1968, pp. 165–184; Also see, Catanach, "Agrarian Disturbance in Nineteenth Century India", *Indian Economic and Social History Review*, Vol. 3, No. 1, March 1966, pp. 66–71; For details about the moneylenders who were assaulted during the Deccan riots see, D. R. Gadgil, "Immigrant Traders in Poona in the Eighteenth Century", *Arth Vignan*, Vol. 1, No. 1, March 1959, p. 16.

their traditional means of livelihood which was jungle produce, also they had to give away their lands which became part of the reserved forests. From being kings of the forests *'jungle ka raja'*, the tribals were reduced to the mere position of forest labourers by the Forest Department and the urban timber merchants and contractors. Following is the testimony in the Bombay Commission Report 1887, of Rao Saheb Ganpatrao Malhar Bokad, a *Mahadev Koli*, aged 50, describing the pathetic situation of the *Kolis*

> "The *Kolis* in Thana are allowed to collect headloads of drywood for sale like the *Thakurs* and others but the supply of drywood is not enough and they are thus driven to manufacture deadwood and run the risk of being punished. ... In Belbihir in Karjat, I know an instance of a *Koli* occupant having sold all the trees in one of his occupied numbers adjoining the forests for Rs. 1.50 for charcoal. He did this because he feared that his land would be taken up for forests. He sold it to a *Talati* of an adjoining village in Murbad. It was a *Talati* who, to induce him to sell, deceived him by telling him that the land would be included in forests. The *Kolis* hold land permanently like other cultivators. Much of the land has passed into the hands of the moneylenders. As a rule the latter are the Kulkarnis who in their dual capacity of village occupants and *savkars* have a powerful hold over the *Kolis*. The *Kolis* therefore can get no help from the *Kulkarnis* and are unable to bring their grievances forward.
>
> I have heard from many people that in Thana, Poona, Ahmednagar and Nasik regular subscriptions are got in the villages twice a year through the village heads to bribe the forest guards, to secure their connivance at getting their ordinary requirements in the forest produce. If more valuable timber is required something extra has to be given by the individual who wants it.
>
> The subscription varies from Re. 1 to Rs. 10 biennially according to the size of the village, either in cash or its equivalent in grain. People do not complain because they think if they did it would lead to worse oppression on the

part of the guard who might take the place of the one they might get punished and removed."[23]

Naturally the discontentment of the *Kolis* was directed towards the *savkars* and petty Forest Officers who played a major role in these state of affairs. Right from the hills of Bhimashankar up to Nasik district all the tribals, mainly *Mahadev Kolis*, *Katkaris* and *Bhils*, got organised against the moneylenders who belonged to the communities like *Marwadis*, *Gujratis*, *Teli*, *Pardesi*, *Brahmins*, *Lingayats* and *Musalmans*. In the area surrounding Junnar *taluka* alone, 300 upper caste moneylenders had taken away all the tribal lands by false means or cheating. This was not the case only in Junnar but also in the regions of Nasik, Dahanu, Javhar, Morwada, Khandesh, Satpura, Igatpuri, Ghoti, Triyambakeshwar, Rajoor, Kotal, Otur, Ambegaon, Khed and *Maval*.

The first revolt against the moneylenders was by Honaji Bhagoji Kingale at Ambegaon *taluka* in *Poona* district in 1874.[24] It continued for a long time in the mountains of Bhimashankar, Ahupe and the *talukas* of Karjat, Ambegaon, Junnar. Finally, Honaji was caught by the police in 1876 and hanged to death.[25] Again in 1889, revolts began in the

---

23 Rao Saheb Ganapatrao Malhar Bokad, Witness No. 19, Oral Evidence, Thana District, *Bombay Forest Commission Report*, Vol. 4, 1887; For economic conditions in the Bombay Presidency during the British rule see, Sumit Guha, *The Agrarian Economy of the Bombay Deccan 1818–1941*, New Delhi, 1985. Also see, R. D. Choksey, *Economic History of Bombay, Deccan and Karnatak (1818–1869)*, Poona, 1945.

24 Case against Honaji Bhagoji Kingale October 1876, Maharashtra State Archives, *Judicial Proceedings*, 1876, 50/1876. 'Honya is the Coeur de Lion with no Saladin to oppose him', were the words of the Under Secretary, describing Honaji Kingale, See J. Nugent, Under Secretary, on W.H. Havelock, Commissioner, Southern Division to Secretary, Judicial Department, No. 763 of 9 April 1874, Bombay Secretariat Record Office, Government of Maharashtra.

25 For more details see, David Hardiman, *Feeding the Banias: Peasants and Usurers in Western India*, Oxford University Press, Delhi, 1996, pp. 234–239.

regions around the Western Ghats of the Bombay Presidency. Vasudev Balawant Phadke was the leader of the revolt near Purandar. The community of Ramoshis in Satara district was led by Umaji Naik. Junnar, Ambegaon in Poona district, Rajoor and Kotool in Nagar district, Goti in Nasik district were the places from where *Mahadev Kolis* organised themselves against the moneylenders and the government.

The famine of 1889 enhanced the misery of the tribals. For want of food and basic necessities, the poor dwellers of the forests had to sell their lands to the moneylenders. They could no longer rely on the forests for their livelihood as they did earlier. This was due to the modern forest policies and the harassment by the Forest Department. Nor could they migrate to some other places as land was no more available free. The discontentment which arose because of all this once again erupted into a revolt. This time the revolt of the *Mahadev Kolis* was led by Rama Kirva who considered Raghoji Bhangre as his idol. Rama Kirva too was hanged by the police. Later, several leaders like Rama More, Sakharam Satpute, Haibati Khade, Balaji Pichad and Dhavala Bhangre organised the people of their own communities and tried to retaliate against the moneylenders and the establishment. Compared to the revolts made by Raghoji Bhangre and Rama Kirva, the other revolts were not very significant though they continued until 1946–47.[26]

Though these revolts were perceived as robberies and anti-social activities by the government, in reality they had a strong support from the local poor who were being exploited by the rich moneylenders by virtue of the new British administrative policies. The poor who were being harassed by the revenue and forest staff, naturally had sympathy for the rebels who had challenged the authority of the *savkars* and Forest Officers. Though these revolts did not have much significance at national level politics, they certainly led to lessening of exploitation of the poor by the money lenders and the forest staff. Most of all it aroused

---

[26] Dr. Govind Gare, *Sahyadriteel Adivasi Mahadev Koli*, (Marathi), Sahadhyayan Publishers, 1974, pp. 67–90; also see, David Hardiman, *Feeding the Baniya : Peasants and Usurers in Western India*, OUP, New Delhi, 1996, pp. 220–234.

pride among the poorest of the lot. According to the report of one British officer, these revolts succeeded in (i) reducing the interest rates on the loans given by the moneylenders, (ii) at least to some extent government became aware of the problems of the tribals, (iii) the behavior of the forest staff towards the tribals improved at least to a certain extent.[27]

During the second half of the nineteenth century an important role was played by national leaders like the members of the Sarvajanik Sabha and Lokmanya Tilak to bring into focus the issue of forest. Tilak through his newspaper began to highlight the problems related to people and the forest. Since the passing of the Forest Act of 1878, news regarding forests was a regular feature in his newspaper. In fact, during the revolts of tribals in 1896, Government suspected involvement of Tilak in instigating the tribals by holding meetings all over the district.[28]

During the 1870s, the Poona Sarvajanik Sabha played a major role in several agitations. The Bombay Presidency witnessed severe economic crisis during the 1870s. Periodic droughts added to the ill effects. The new land revenue settlement of the British created havoc in rural Maharashtra. Rural indebtedness rose inexorably leading to several tensions between money lenders and peasants, which finally exploded into the Deccan riots. From 1876 to 1879, western Maharashtra suffered the worst famine of the century. Problems regarding food supplies, water supplies, increasing forest restrictions, increasing prices of essential commodities and so on mounted to great heights. However, the government was stuck with its policy of fulfillment of long-term objectives. During such a period the Poona Sarvajanik Sabha worked relentlessly for the people. A network of school teachers, retired civil servants and others organised the people and were more efficient than the employees of the government.[29]

---

27 Dr. Govind Gare, *Sahyadriteel Adivasi Mahadev Koli* (Marathi), Sahadhyayan Publishers, 1974, pp. 67–90.

28 *Files of the Revenue Department*, Vol. 30, 1898, p. 69.

29 *Report of the Poona Sarvajanik Sabha* submitted to the East India Finance Committee, Poona, 1873; also see, Ravindra Kumar, "Deccan Riots of 1875", *Journal of Asian Studies*, Vol. XXIV, August 1965, pp. 613–635; also see,

Poona Sarvajanik Sabha was originally founded in 1867 by the *Brahmins* of Poona. This organisation was to serve the dual objective of educating the public on the crucial issues of the day and communicating the views of the people to the British Government. In 1870, it was reorganised along more ambitious lines as the Poona Sarvajanik Sabha. This Sabha claimed sympathy of traditional aristocrats like Pant Pratinidhi of Oudh and Maharaja of Kolhapur, who gave the legitimacy essential to the Sabha. The most sophisticated political thinker of his time in Western India M. G. Ranade and the liberal intellectual Ganesh Vasudev Joshi were among its members. These Western-educated elite members shaped the political activity of this Sabha.

At the time of its foundation the Sabha consisted of 95 representatives from all over Maharashtra who attended the inaugural meeting on 2 April 1870. These representatives were elected by over 6,000 persons representing all castes, creeds and interests.[30]

Though this Sabha was dominated by one particular caste of *Chitpavan Brahmins* who belonged to the elitist strata of society, it attempted to bridge the gap between the peasants and the State by performing the job of the intermediary. Members like M. G. Ranade were eager to bring forward the problems of peasants and build a social base for the Sabha in rural areas.

On the resettlement of Deccan by Francis, Superintendent of 'The Revenue Survey and Assessment of the Deccan' considerable opposition was expressed by the peasants. At this juncture Poona Sarvajanik Sabha came forward and formed a sub-committee which presented petitions to the government about the grievances of the

---

Richard Tucker, "Forest Management and Imperial Politics in Thana District, Bombay, 1823–1887", *Indian Economic and Social History Review*, Vol. XVI, No. 3, 1979, pp. 273–299.

30 "Memorandum on the founding of the Poona Sarvajanik Sabha, Unpublished collections of Bombay State Committee for History of Freedom Movement in India", cited in Ravindra Kumar, *Western India in the Nineteenth Century: A Study in the Social History of Maharashtra*, Routledge and Kegan Paul, London, 1968, pp. 613–635.

peasants. It also dispatched trained cadres to the villages in order to acquaint the peasants with reasons for their miserable plight. Hence in the words of Ravindra Kumar they emerged to be "An elite which combined traditional techniques of agitation with a commitment to Western political ideals, presented a serious threat to the British authority and the significance of this development was not lost on the British Government."[31]

The report of the Sub-Committee on Deccan, in the words of Ravindra Kumar, "...Though written contemporaneously, the petition and report stood world apart in intellectual attitudes. They displayed the cleavage in Maharashtra between those who subscribed to traditional style of politics and the emerging Westernised elite which was attempting to secure positions of leadership within the community. There can be no doubt of the greater impact the agitation of the Sabha had on government as well as on the rural community. For the leaders of the Sabha understood the political beliefs of the new rulers, and could use techniques of agitation which the *Kunbis* understood."[32]

In the report to the East India Finance Committee on matters relating to India the Sub-Committee stated "...The Forest Department, though latest born of the departments, in Thana district, especially, has been exacting and oppressive to a degree unequalled by any other department of revenue. Not only have the guaranteed rights of private proprietors in the forests preserved by them been interfered with, but all manner of exaction and levies of a miscellaneous kind have been made without any legal authority."[33]

According to the report, many suits were filed against the Forest Department by the villagers in most of which the acts of the officers were condemned by the judges and the government was saddled with

---

[31] *Ibid*, p. 628.

[32] *Ibid*, p. 627; also quoted by Richard Tucker, "Forest Management and Imperial Politics," *The Indian Economic and Social History Review*, Vol. XVI, No. 3.

[33] *Report from the Sub-committee of Poona Sarvajanik Sabha* appointed to collect information to be laid before the East India Finance Committee, Poona, 1873, p. 77.

the costs. However, only a few had the courage and means to fight their disputes in civil courts. Hence the illegal oppressions in the larger number of cases continued to exist with more or less force.

One of the complaints mentioned in this report was that by the custom of the country and guarantee of government, when a rate had been once fixed upon revenue paying land, firewood and timber trees which the *rayots* planted in their own lands like other produce belonged to him. The British Government had always professed to act on this understanding (vide proclamation No. 108 of 1847 and Bombay Government Resolutions No. 49 of 1847 and No. 1240 of 1868, as also Joint Rules Clauses 1 and 10). However, government officers in the Forest Department had begun to show disregard to these guarantees and disputed the *rayots* claim of property in his own trees.[34]

Another complaint was that in the year 1869–1870, the right to remove drywood from government forests was sold by auction. The auction purchaser was subsequently, by an order of the Deputy Conservator of Forest, permitted to remove drywood from the *rayot's* lands. This was found very oppressive by the people and a memorial was sent to the government. A local newspaper of Thana had also taken up the issue, but no notice was taken by the government and the oppression continued unchecked. Besides this, the *rayots* claimed the

---

[34] *Ibid*, p. 78. In case of Thana district, prohibition to cut timber growing in the *rayot's* own lands extended to teakwood and blackwood only. In spite of this, prohibition was later extended to every variety of timber trees growing on the *rayot's* land. This subsequent circular was not explained nor published for the information of the people. Due to this the Mamlatdar of Dahanu was charged with having negligently caused loss to the government, by allowing the people to cut their own trees against the order contained in the unpublished circular. However, after an inquiry the Mamlatdar was proved innocent and the Circular order No. 100, dated 6 March 1866, affirmed the *rayot's* right on his own trees by declaring that if any person wanted to cut down timber growing in land not his own, permission would be given to him with the consent of the owner of the number. The people were worried that this circular could not be repealed by the unpublished private circular issued by the Survey Department, mentioned above.

trees growing on their land by arguing that it took 40 years for timber trees to grow to a condition fit for cutting. During all those years, the owner bore the loss of the ground under the trees planted by him and paid full government assessment and yet when the trees grew, the government stepped in and prohibited the owner from deriving the full benefit of his investment, which they found was unjust.

Yet another complaint noted by this Sub-Committee was that in the year 1863–1864 Mr. Ellis, Revenue Commissioner, had issued an order that in the case of those private forests from which it was customary to exact a share of the produce of the trees growing on them, the government would give up its rights forever to interfere with the owners of such lands on receipt from them of 10 years' value of the estimated produce.

Nevertheless, 2 years after when the government had received in full the commutation value of the 10 years' produce for the forest lands which were private property of the *rayots*, far from ceasing to interfere any further with the numbers so settled, Forest Officers made it essential for the *rayots* to acquire permission to cut timber from trees growing on their own land. Consequently, a suit was filed against the government, the judgment of which came in favour of the people. Notwithstanding the result of this suit, the government officers continued to interfere with the *rayot's* management of lands settled under the 10 years commutation rule, due to which many people felt oppressed.[35]

Discontentment among the people over forest issues continued to increase, as several new restrictions were imposed on the use of forests by them with the passing of the Forest Act of 1878. Even after closing the reserved areas, the forest area which was meant for the use of people was not totally free in the true sense of the word. People's movements in these forests were also restricted. They could avail of the facility granted by the Forest Department only at certain times. Also, this could be done only if people had permits and if they paid the grazing fees. The regulation to prevent forest fire was the most

---

[35] *Ibid*, pp. 79–80.

controversial as it was difficult to discover who had set fire. The department not only required villagers to help to fight fires but they were held collectively responsible for fires if the guilty were not found. This issue continued to bother people even in the first half of the twentieth century.[36]

Poor peasants and tribals though discontented were in no position to protest in an organised manner. Individual offence was one form they could adhere to. Due to this the number of forest offences increased to such an extent that the Forest Department had begun to be identified as the Police Department by the people. Bribery and corruption among forest officials at the lower level became a common phenomenon. The most notorious practice was a blackmail system, in which forest guards threatened poor villagers with fines or imprisonment for imaginary violations of new regulations. The government cautioned the Forest Department to pay due regards to the habits and wants of the poorest class of the population, knowing well, that this class resented vexatious and oppressive interference with their daily life. However, the Forest Department did not pay heed to it.[37]

Poona Sarvajanik Sabha time and again kept informing the government about the people's grievances over the forest issue. Through its journal, the Sabha agreed in principle over the need to conserve the forests. However, it also pointed out that the policy of the Revenue Department to bring more and more land under cultivation in order to meet the enhanced revenue requirements of the government was grossly responsible for the destruction of forests. The Sabha agreed that more severe regulations might be necessary and appropriate in reserved forests and for cutting of commercial timber. For all other forest produce including grazing rights, it believed, there was no need for any prohibition or detailed system of fees and permits. However, the Sabha pointed out that the natives of the country were

---

[36] *Forest Grievance Committee Report*, Bombay, 1927.

[37] Standing Orders of the Bombay Presidency, corrected upto 1910, Government Press, Bombay, 1924, pp. 191–236.

better qualified to look after the forests because of their instinctive and religious reverence for trees and the vegetable kingdom than any officer of the dominant race.[38]

Thana Forest Association was another organisation, which worked in alliance with the Poona Sarvajanik Sabha to represent people's grievances over the forest matters during this period local landlords were its members.[39] During this period a memorial was sent by 5,000 inhabitants of Kolaba against the Indian Forest Bill, dated 21 December 1877. The memorial stated

> "Inhabitants fear that if the Forest Bill (1878) be passed into the law, the rights, wages and privileges relating to the forests which they have enjoyed from time immemorial will be violated seriously. They therefore, humbly make some suggestions for consideration at the time of passing the bill. The bill is to be made informally applicable to the whole of British India.
>
> In Konkan no crops can be raised on the greater part of its lands, without the lands being burnt every year. And thus, there is an inseparable connection between the raising of the crops and the cutting of the jungle in the forests. That is to say, the grass, the brushwood and the small branches of trees and twigs are absolutely necessary for the purpose of burning the soil. This shows that even though the proposed law prohibiting all access to the forests, may perhaps be applicable to some parts of India, it will certainly be mischievous and pernicious in its effects to the country. That from the time immemorial, before the advent of the British rule and since the introduction of this rule upto this day, it has been the undisputed practice of the people to make use of all the trees in the forests adjoining to their

38 *Quarterly Journal of Poona Sarvajanik Sabha*, December 1882–1883.

39 *Bombay Forest Committee*, Vol. 1, 1887; also see, Indra Saldhana, "Colonial Forest Regulations", Richard Grove, Vignette Damodaran and Staple Sangwan, (ed.), *Nature and the Orient*, OUP, New Delhi, 1998.

villages for the purpose of making cuttings for burning their land with, for raising their crops therein, and this practice was confirmed even at the introduction of the revenue survey. Before the introduction of survey, only cultivable lands were liable to the government assessment and the payment of the assessment gave only the *rayots* the right of making cuttings in the forests of their villages and territory, without any extra payment for the use of the jungle. In the survey all lands, cultivable and non-cultivable such as grass and brushwood lands and "c" were measured and surveyed and rendered liable to assessment. But at that time, the most distinct and clear promises were given by the government to the people that they would not be subjected to any extra payment for the use of the jungle or for other articles, such as stone, sand and "c". Yet the Government of Bombay, passed an Act No. 3 of 1869 and legalised a tax called the local funds cess imposed on them before the passing of the Act. The same was accepted by the people through feelings of loyalty. But the government did not then interfere with and disturb peoples' immemorial right of making use of the forests for their agricultural and horticultural purposes. Till today the people have been allowed to cut all kinds of trees growing in their proprietary and private revenue survey numbers and all trees, except teak and sissu trees, growing in the public forests. This usage has continued undisturbed even under the British Government for more than 30 years and under the native governments from time immemorial. Thus, the people have acquired prescriptive rights to use of the jungle. Yet the proposal law seeks to destroy these rights and violate all the given promises at one blow.

People submit that even though the peasants pay the full assessment on their lands, yet they are not to be allowed for their agricultural operations, it will not only be no encouragement to agriculture but will be a positive discouragement and impediment to their occupation, nay it will be an obstruction worse than any which a most despotic government may devise for the annoyance of its subjects.

Mr. Dunlop, the Collector of Konkan, was the first to make rules for the regulation of the forests and published a proclamation on the 1 March, 1824. Dr. Gibson was the second officer who made any rules on this matter. But both these officers reserved full liberties to the people to make use of the forests for their agricultural and horticultural purposes. But the proposed legislation seeked to place entire restrictions on the use of the forests. People requested that if the promises given by such high government officers, as those named above are now violated by special legislation, the *rayots* will not rely on the promises of the government.

People requested that the present legislation by placing entire restrictions on the use of the jungle, will be highly injurious to the peaceful inhabitants of the country not perhaps the Supreme Government imagined by. Wild people like the Katkaris, Thakurs, Kolis etc. live in the forests of the country. The corn they grow for their use in the hilly forests is grown without the lands being ploughed beforehand. Hence, the yields of their crop is barely sufficient for their maintenance even for a few weeks in a year. Their principle means of livelihood therefore, is the cutting of drywood in the forests and selling the same for fuel to the inhabitants of the villages. This has been their occupation and mode of earning livelihood for ages past. But if this is now stopped, it will be extremely difficult for them to earn their maintenance without a resort to unlawful means, such as plundering the armless and helpless inhabitants of the villages. Yet the present bill, proposes to treat their occupation as a penal offence: the result of which will be, if the act will be enforced, that these classes of wild people will surely plunder the people of the villages. In addition to this, in such a case thousands of these wild people will have to be thrown into the jails and maintained at the cost of the general public.

> That the majority of the people of this country living in the villages, are in a very poor condition indeed. Agriculture is the only means of earning their livelihood. But the yield of their lands is so small, that they are seldom able to get full nourishment for more than 2 months in a year. The rest of their agricultural produce goes to landlords, the government assessment and the demands of their creditors. For 10 months in the year, therefore, they have to maintain themselves by hard labour. One of these occupations is to cut drywood from the forests and to sell it. By the present legislation they are to be deprived of even the precarious mode of earning livelihood, for even though they work hard at great risk to their lives, they have to live upon only one fourth the quality of nourishing food which is necessary for a man. Three fourth of the substances they eat consist of simple vegetables and roots. Yet, such a wrested class of people will be deprived of an occupation. Besides they will have to pay for their fuel which from time immemorial they have been obtaining free of payment by the labour of their own hands.
>
> People requested to allow the liberties and rights in regard to the use of the forests which they have enjoyed from immemorial time."[40]

In 1882, when the forest regulations were extended to cover uncultivated *Gurucharans* and village wastelands on which villagers had previously depended for their needs of forest produce, the discontent intensified. It reached such an extent that in 1885 a deputation consisting of influential people from Thana and Kalyan along with eminent personalities like Kashinath Trimbak Telang, Dadabhai Naoroji, P. M. Mehta and others met the Viceroy to acquaint him with the hardships caused by the Act. It asked for temporary suspension of the forest rules and for appointment of a commission of

---

40 *Forest Branch Proceedings for March, No. 54, Part B,* Department of Revenue and Agriculture, Government of India, National Archives of India, New Delhi.

inquiry to look into the situation.[41] Consequently, the Government of India in Calcutta announced the formation of a special commission, The Bombay Forest Enquiry Commission to study the situation in Thana and Kolaba districts. Its members were four British officials representing the Revenue and Forest Departments and three Indians including K. L. Nulkar who was the Chairman of the Managing Committee of Poona Sarvajanik Sabha.[42]

By this time, even the media in the Bombay Presidency had become aware of the forest problems faced by the rural population. It began giving extensive coverage to this issue. The Native Opinion of Bombay argued that the 'industries of the people' were being destroyed by excessive forest regulations.[43]

The leading newspaper of Thana, *Arunodaya*, gave a vivid description of the local conditions and insisted that the peasants, untouchables and tribals required free access to timber without any powers to the *Mamlatdar* to regulate the system.[44]

Kesari, the militant nationalist newspaper, which was run by the most prominent militant leader of Bombay Presidency – Bal Gangadhar Tilak, took a strong position over this issue. Even after the death of this leader, forests continued to be a major concern of this newspaper until independence. *Kesari* editorials put forth the concept of local control where the village officials were vested with the power to control the wastelands. Rural District Boards rather than the Revenue Department were to supervise the work of the Forest Department.[45]

---

41 *Revenue Department*, Vol. 105 A, 1885.

42 Mr. Nulkar took fundamental exception to major element of the report. See *Bombay Forest Commission Report*, 1887, Bombay.

43 Richard Tucker, "Forest Management and Imperial Politics", *The Indian Economic and Social History Review*, Volume XVI, No. 3, Bombay Presidency, Reports on the Native Papers, 11 July 1885, pp. 5.

44 In the memorials submitted to the Commission one was submitted by 4 inhabitants of Thana, of whom one was the Editor of *Arunodaya*, the remaining three being the lawyers; See *Bombay Forest Commission Report*, Vol. I, 1887.

45 Richard Tucker, "Forest Management and Imperial Politics", *IESHR*, Vol. XVI, No. 3.

Though *Kesari* recognised the importance of having a Forest Department in order to protect forests, it did not appreciate the way in which people's age old rights over forests were denied to them by the Forest Department. It held that the greed of the Forest Department to make money through forest produce was the most important cause of all the trouble. Also, *Kesari* argued that in case of self government, even if there was injustice it was not felt as badly as that of the injustice by foreign rulers. Largely because the income earned by the government of foreigners like the British went to their mother country. It did not remain nor was it spent in the native country.[46] The newspaper welcomed the appointment of the Forest Commission, especially the fact that a few native members were also appointed to the Commission. It helped to make people aware of the appointment of the Commission. It published the questionnaires distributed by the Commission and discussed the problems of the *rayots* in detail and offered valuable suggestions.

Arunodaya desired that the local wood depots be run by Citizen's Committees elected by local populations and not by Forest Officers.[47] The Forest Association of Thana district expressed its views in its testimony to the Forest Commission. It asserted that the British regime had claimed a right to own or regulate forest produce before 1862. However, in the 16 years, since the formal establishment of the Forest Department, it had gradually enhanced its revenue earning capacity and begun to curtail people's rights. The Association accused the Forest Department of perpetrating a sudden and violent confiscation of the long cherished traditional rights of the people since 1878.[48] The Association advised the department to limit its role to purely technical and advisory matters. It was desired that the department should advise only *taluka* boards, which already had significant representation from men of local influence and had become a genuine representative system of local government. Complete control over communal forests by the board was considered advisable as only this board, they believed, could

46 *Kesari*, 9 June 1885.

47 *Bombay Forest Commission Report*, 1887, Vol. 1, pp. 63–77.

48 *Ibid*.

gain people's confidence. The board was expected to curtail the power of outside merchants by working with the department to channel all shipment of timber through government run depots.

It was largely the interests of the peasants, which were given priority by the Indian representatives. Though the representatives largely belonged to the elitist *Chitpavan Brahmins*, their sympathy was with the *rayots*. Hailing from the land-owning class of the coastal districts of the Bombay Presidency, they were able to appreciate the problems of the agriculturists in a better manner. These representatives of the agricultural class naturally resented the interests of the class of urban merchants and contractors. Urban timber contractors were recognised as those who were not in the habit of always keeping themselves within the limits of their contract terms.[49]

Some memorials argued openly that local villages should be given full priority over all outsiders, whether British officials or urban Indian interests. The testimony offered by witnesses in front of the Commission gives important insight into the intricate problems faced by the locals, which were specific to their traditional professions and lifestyles that were affected due to the new forest regulations. For example, Witness No. 3 stated that the poorest of the lot excepting the *Pandharpesha*, that is the middle class income group, lived more upon *Varkas* grains like *nagli* than on rice. They generally grew enough *Varkas* for their own consumption. These grains though less palatable were considered more nutritious and filling. If the supply of these grains was not enough, they sold paddy to buy *nagli* and the like. However, the crops like *nagli* and other grains required *rab* which had become impossible due to the modern regulations.[50]

Jagu Govind Patil, a Kunbi policeman of Mosoli village in Thana district said "Most of the *rab* beds earlier were made with cow dung. The reduction of the cattle in later years is due to contraction of grazing

---

49 Richard Tucker, "Forest Management and Imperial Politics: Thana district, Bombay, 1823–1887", *IESHR*, Vol. 26, No. 3, pp. 291–293.

50 Witness No. 3, *Bombay Forest Commission Report*, Vol. IV, 1887, Thana District.

area on account of breaking up of land for cultivation at the survey and stricter forest conservancy...Until 7 or 8 years ago, fuel was freely permitted to be removed from the forests of Kainad and Saravat and the people had no occasion to use cow dung for fuel. But since the imposition of forest restrictions, people use part of their cow dung as fuel. This proportionately reduces the supply of *rab*. The export of grass commenced only 3 or 4 years ago. This export is an additional reason for the decrease in cattle that is going on year after year."[51] This transition of cultivators from fuelwood to cow dung for the purpose of fuel has been pointed out by T. N. Atre in *Gao-gada*. He has also described how due to the scarcity of fuelwood, people had to turn from fuelwood to cow dung for burning. This in turn, reduced the manure for the fields which was a consequence of the forest restrictions.[52]

Another witness Babuji Krishnaji, a Maratha cultivator, aged 70 years, residing at Lonivli in Panvel *taluka* of Kolaba district said

> "...It has been customary with us to get tahal from the jungle...The loppings of all trees except teak and blackwood are removed for *rab*. Ain, nana, bondga, dhavda and hed are considered the best materials for *rab*, while ukshi, dhayati, karvand, phangli and kuda are considered as inferior *rab* materials. About 20 years ago, I could not manage to burn the seed beds with *rab*. I therefore, on two occasions raised seedlings by manuring the plot with oil cake. The result was indifferent seedlings and a poor crop.

---

51 *Ibid*.

52 T. N. Tare, *Gao-gada*, Mote Publications, Bombay, 1915. In fact, J.A. Voelcker too recommended very strongly the need to make provision for fuel for the cultivators. He said, "I regard the provision of fuel as the most potent means of maintaining prosperity not alone to the cultivators but to the state itself... If wood could be made to take the place of dung for fuel we should soon come to realise that more wood means more manure, that more manure means more crops and more crops an increasing revenue to the state." For details see, J.A. Voelcker, *Report on the Improvement of Indian Agriculture*, Government Press, Calcutta 1893, p. 136.

> Where I ordinarily got one candy of paddy, I could get only 5–7 maunds. The cost of manuring the seed bed with oil cake for one acre of rice field is about Rs 5. I have come to the conclusion that tahal *rab* is essential for the rice cultivation in this district...About 20 years ago, I had about 80 heads of cattle. They gradually got reduced. I do not buy new cattle on account of difficulties of grazing created by occupying more area for cultivation and conservation of forests. When I had the full number of cattle I could get sufficient cow dung to burn the *rab* beds for 10 or 12 acres of kharif land."[53]

Prohibition to collect *rab* was one of the most common complaints of the people. Whereas *rab* practice was one of the most hated practices from the point of view of the Forest Officers. However, the agricultural experts held a different view from that of the Forest Officers on the usefulness of *rab*. E. C. Ozanne, Director of Agriculture, clearly stated that, "...In Thana, loppings are absolutely necessary because other *rab* material is not sufficient to meet the demand for rice, *nagli* and *varai*."[54]

In actuality, the important factor responsible for degradation of forests was the transfer of rice lands without the *shindad* land, to which they were by custom attached and from which the raw material

---

53 *Bombay Forest Commission Report*, Vol. IV, 1887, Witness No. 25, "Oral Evidence", Kolaba district.

54 *Ibid*, Vol. 4, p. 232. E. C. Ozanne observed that "the necessity of *rab* in the Konkan was to be appreciated with reference to two essential features: 1. the peculiarity of the rain on the Western coast, its intensity and continuity in early monsoon and also in early cessation; and 2. lack of water storage facilities in the Konkan". J. A. Voelcker while commenting on the lopping of trees for the purpose of *rab*, stated that "the shoots of some type of the lopped trees take 3 years to grow to a sufficient size and the trees live for 40 years. I could not help thinking it was much better for the trees to thus, yield a triennial supply of shoots for 40 years than that they should be left alone all the time, in order to afford at the close of it one single log of timber." p. 145.

required for cultivation was obtained.[55] The introduction of survey settlement in these tracts had resulted in the conversion of *varkas* land into separate occupancies transferable at the will of the occupant with or without the rice land, to which they were formerly attached.[56]

Similar complaints were made by several people from both Kolaba and Thana district. For centuries together, they had been practicing *rab* cultivation and suddenly, they were stopped from doing it. Also, many a times, the Forest Department while acquiring land for reserved forests gave some other pieces of land in exchange to the people. However, these lands were not of as good quality as their own lands. Also, in terms of distance they were not very convenient. So one can see lack of understanding of the rural conditions on the part of the Forest Officers. Witness No. 10 of Mahad *taluka*, Gulam Valad Abdul Rahman, a *Musalman* cultivator and trader aged 45 years said "...The condition of the cultivators of my *taluka* is depressed. They have little or no credit and the *savkars* refuse to advance them anything, unless they execute bonds on duly stamped papers. The cultivation of *nachani* and *vari* has decreased because the cultivation in forests has been stopped and *rab* material cannot be had so easily as before. In the village of Kamble, where 50 candies of *nagli* were grown, 5 candies of barley are produced now, and instead of being an exporting village it has to buy other markets... We were offered lands in exchange in the village of Nargaon, but we refused to take it as it was unfit for cultivation. Our lands are superior; we can grow pulses in them."[57]

---

55 The inseparability of rice lands from *shindad* lands was a special feature of the *khasbandi* tenure formerly prevailing in parts of the Konkan. Indra Munshi Saldhana, "The Political Ecology of Traditional Farming Practices in Thana district, Maharashtra (India)", *The Journal of Peasant Studies*, Vol. 17, No. 3, 1990, pp. 433–443.

56 *Bombay Forest Commission*, Vol. IV, 1887, p. 101; also see *Kesari*, 21 April 1885.

57 Witness No. 10, Kolaba district, *Bombay Forest Commission*, Vol. IV, 1887; there were several instances where land was taken away from the people without any compensation. This deprivation at times led such people to unauthorised cultivation. For instance Orr, the Collector of Thana while

Witness No. 19 Pandurang Laxman Dandekar, though a *Brahman* moneylender reported... "The occupants of lands are required to obtain passes before cutting wood from their numbers. This in itself is a cause of distrust created in the minds of the occupants with regard to their rights...The condition of people in Mahad, Roha and Mangaon is a depressed one. The *Katkaris* and *Thakurs* are not allowed to cultivate *dalhi*, nor do they get employment under the wood contractors. They are thus, in a much more worse condition than before. The agriculturists are not so well off as before, because they do not get the wood and *rab* they require, so their lands produce less. The wild tribes resort to thieving as they cannot get labour."[58]

Though *rab* cultivation was made into a big issue by the British, people did not cut the *rab* (lopping the branches) indiscriminately. There was some kind of discipline and systematic method according to which these operations were undertaken. As Witness No. 7[59] stated "The *Phasepardhis*, a hunting caste among tribals, stored sufficient fuel in April or May to last for the entire monsoon. One even stored a year's

---

handing over the charge to Wood wrote in a note that ..."There are tracts where it (unauthorised cultivation) needs lenient treatment. People have been deprived without compensation of lands which they have habitually tilled...I believe there are cases where the deprivation has been a mistake and ought to be abandoned..." See, *Forest Administration of the Bombay Presidency 1901–1902*, p. 23.

58 *Ibid*, Witness No. 10. The testimony of the witness No. 10 mentioned above, points the main weakness of the Forest Department which was demoralisation of the people, due to prohibition. It shows the misery caused by the department which led to the poorest of the lot into committing so-called forest offenses. This situation is comparable to the 'right to subsistence' as described by James Scott. Right to subsistence, according to James Scott, is one of the primary moral expectations of peasants, constituting 'the moral economy of the poor,' a concept adapted from E. P. Thompson's writings. See for further details, James Scott, *The Moral Economy of the Peasant : Rebellion and Subsistence in Southeast Asia*, New Heaven, 1976, and E. P. Thompson, "The Moral Economy of the English Crowd in the Eighteenth Century", *Past and Present*, Vol. 50, 1971, p. 50.

59 *Ibid*.

supply at a time. For the remaining part of the year, they brought only as much fuel as to meet their immediate requirements. The tendency to accumulate, hoard and sell was not known to them." Also, a date was fixed for the beginning of the collection of *tahal* (lopping and collecting of branches and deadwood for *rab* cultivation) and cattle dung for all the neighbouring villages. Before this date, nobody was permitted to collect even one stick for *tahal* or cattle dung. The process was always supposed to begin with a ceremony of breaking a coconut. This ritual probably was meant to offer an equal opportunity to all to collect the necessary *tahal* and cattle dung. This only proves that even with the old traditions, people were not allowed to fell trees or its branches indiscriminately. However, the British policy makers alien to this culture were not able to understand it." [60] In fact, they always took it for granted that people were wasteful in nature and that "they took all that they required for their simple wants, where they found it."[61]

Many a times, the tribals were turned out of their hamlets by the Forest Department in order to secure reserved forests. Though the tribals were offered compensation in terms of money, it was nowhere close to what they lost in terms of cattle and members of their own community, apart from the land. The land that was offered to them for cultivation as an alternative was of an inferior sort which was not liked by them, and hence they refused it.[62] People also felt cheated when they saw the government cutting trees to supply to Mumbai while the local villagers were prohibited from doing so.[63]

Kashinath Vishnu Phadke, a *Brahman* cultivator and moneylender aged 50 residing in Thana, stated that "...Comparing the state I found the forests then into their present state, I am of the opinion that the forests are much poorer, that is to say, they are not so many large-

---

[60] *Ibid*, Shivaji More's testimony, Witness No. 5.

[61] B. Ribbenstrop, *Forests in British India*, First print 1900, Delhi reprint 1986, p. 66.

[62] *Bombay Forest Commission Report*, Shivaji More's testimony, Witness No. 5, Kolaba district.

[63] *Ibid*, Second Witness, Kolaba district.

sized trees as were to be seen before. I was employed as a measurer about the same time in Uran and Garoli, and I can say the same thing about the range of forests which run from Ghodbundar to Ḳurla. I think, the chief reasons of this deterioration is the restrictions imposed by the government. This causes waste. A man to get a plough goes and clandestinely cuts a whole tree, takes the part for the plough and leaves the rest which simply goes as kolpat (material used for burning and preparation of the seedbed before souring. The ash generated due to burning of this material acts as nutrient for the soil.), whereas formerly the whole tree would have been utilised as far as possible and would have supplied other agricultural wants. Strict conservancy has demoralised the people and they do not think it worthwhile to protect the saplings as they used to do. Another reason is that government tries to make money out of the forests by cutting wood...If the people were guaranteed the full and unreserved right to all the trees growing on their lands, they would be encouraged to plant more trees."[64]

Bapu Dadoji Korde, government pleader, aged 52, residing at Morbad, said

> "...I do not know of any land specially set apart as forests at the time of the survey. 7 or 8 years afterwards some lands were set apart as forests in about 25 villages... The quantity of forest growth is very much less now than it was before 1852–53. Between this period and 1862–63 government gave contracts for felling timber for sleepers, building purposes and to a large extent to traders, and hence the present diminished supply. The contractors cut timber fraudulently from *gurucharan* and *malki* lands as well as government forests taking advantage of the fact that the latter were not clearly marked off from other lands. Teak

---

64 *Ibid*, Witness No. 6, Kolaba district. Similar type of destruction of forests for agrarian extension was observed by the Superintendent, Deccan and Gujarat Revenue Service, see, *Papers relating to the revision survey settlement of the Shahada taluka of the Khandesh Collectorate*, Selections from Government Records, No. 378, New Series, Mumbai, 1899, p. 11, Maharashtra Archives.

> has increased but *injali* or inferior wood cannot be got off so good quality as before.
>
> Another reason for denudation of occupied lands is that the owners themselves cut trees, thinking that as government contractors had already in some cases cut trees under cover of their contracts through fraud or ignorance in occupied lands, it would be better to avoid the chance of such an event occurring again to cut the trees and realise the value themselves. People also thought that as government had prohibited the cutting of teak they might also prohibit the cutting of other trees.
>
> Our claim is to be allowed grass and *tahal* for cultivation, firewood, wood for agricultural purposes, wood for building common houses, material for fences, medicinal products from the forests such as honey, *hirda*, wax, *beheda*, *avala*, marking nuts and minor fruits and flowers, *apta palas*, *tembhurni* and teak leaves. Some *Kunbis* would also want timber for fish dams."[65]

This testimony reveals that people were well aware of what was being taken away from them and what they were being deprived of. The injustice that was being done was obvious even to the poorest of the illiterate tribals. Due to this those who had trees in their own lands hastened their cutting. They started felling even Mowra and fruit trees, afraid that they would not be anyway able to use the fruits in future. In this context a popular anecdote emerged, which succinctly revealed

---

[65] *Ibid*, Witness No. 10, Kolaba district; It was not only the small land holders who started felling trees indiscriminately with the fear, that if they do not utilise their own trees sooner or later the government merchants and contractors would take them away, but also the small rulers and big landlords and zamindars hastened the process of exploitation of their own forests due to the same fear. See, Mark Poffenberger and Chatrapati Singh, "Communities and the State: Re-establishing the Balance in Indian Forest Policy", *Village Voices, Forest Choices : Joint Forest Management in India*, (ed.), Mark Poffenberger and Betsy McGean, Oxford University Press, Delhi, 1996, p. 60.

the situation prevailing then. People said "who knows whether the government would not prohibit us to cut math."[66]

Some of them protested vehemently, saying that if their conserved land was to be included in reserved forests, they would consider it robbed. They stated "we are to look upon the government as *mai baap* (parents) and ask from them trees although they are ours. No permission should be required to be obtained for cutting wood in *malki* lands for agricultural and domestic purposes. Before the *rayots* can be induced to assist increasing and conserving the forests, they will have to be allowed the same liberty that they formerly enjoyed in the matter of taking forest produce. If this be done the *rayots* will cordially assist the forest conservancy. But as long as this is not done, any addition to the forest establishment, even to the extent of employing one man to look after each tree will not mend matters."[67]

However, to expect the forest policy makers to understand people's needs was not realistic as they did not hesitate to bring even burial grounds under reserve forests. Witness Pandurang Hari, a *Brahmin vakil* in Subordinate Judges Court reported "...I know difficulties occur in procuring wood for cremation owing to forest restrictions. The people experienced the greatest difficulty in this matter in 1882. In one case, a man had to pull down a part of his house to burn a dead body and in 3 instances within my knowledge people had to bury dead bodies for want of fuel."[68] Such was the gravity of peoples' problems, which had occurred due to the forest restrictions.

Harassment by forest staff was another major problem reported by witnesses in front of the commission. Bachubin Ram Patil, a *Maratha* cultivator aged 50 from *taluka* Khadki said "We obtain our fuel and wood for agricultural implements from our forests not withstanding the prohibition because we cannot help to do so. There is an agricultural depot established at Bhiwandi which is 8 miles from my village. There are no other depots at shorter distance. People can not afford to purchase wood from such depots. None of the people from my village

---

66 *Ibid*, Witness No. 3, Thana district. (Math is a common pot herb.)

67 *Ibid*.

68 *Ibid*, Witness No. 40, Thana District.

buy any wood from Bhiwandi. They manage to get wood from their village forests by eluding the vigilance of the forest guards. The people feel they are at the mercy of the forest guards, who are looked upon as Gods. The forests were taken care of before by the village officers and the *rayots*. The forest guards are only kept up to plunder the people."[69]

The following evidence supports this complaint. Rao Saheb Bhao Ramchandra, a *Prabhu Mamlatdar* of Kajan village stated "...Lately the Forest Department has become very unpopular because the number of low paid subordinates has largely increased. They do not patrol the forests, but are constantly worrying the people in the villages by searching their houses etc. in quest of offences and when *tahal* and other produce are found with them, the same are often attached on suspicion and the subsequent enquiry causes much annoyance and loss of time to the people. Hence, the discontent. Poor people who confermed to the rules and went through the procedure of getting permits would have to pay, besides the 1 anna for the permit, 2 *annas* more in the shape of lost time and labour. Therefore, they get firewood without permits and if caught get into trouble. So they have to keep in good terms with the guards"[70].

"...If the village officers look after the forests there is no need for such a large subordinate forest establishment."[70] However, harassment by lower level forest officials was a common complaint which continued to exist even in the first half of the twentieth century. It was also one of the causes of the many revolts which had taken place since the

[69] *Ibid*, Corruption among the lower level forest staff had become a common phenomenon during the British rule. Inspite of the several protests made by the people against the conduct of the forest guards, there was no improvement in the situation which is obvious from several administrative reports of the Forest Department, for example, the *Administrative Report of 1902–03*, where two Beat Guards were prosecuted before the First Class Magistrate, one for misappropriating property and the other for extorting money by threats. See, *Forest Administrative Report of the Bombay Presidency*, Chapter 4, 1901–02, p. 22; This situation continued to exist even in the latter period. See, Godavari Parulekar, *Jenvha Manus Jaga Hoto*, (Marathi), Mouz Publications, Bombay, 1969.

[70] *Ibid*, Witness No. 6, Thana District.

formation of the Forest Departments.[71] *Kesari,* the Marathi daily, highlighted this issue in its issue of 18 October 1892. When the Government started blaming the lower level forest officials for harassing people, *Kesari* pointed out that if the act by itself was unjust, there was no use blaming the lower level forest staff for it. It further suggested the need to make some amendments in the act.[72]

In their complaint Mr. Kashinath Vishnu Phadke, Chintaman Sakharam Karandikar, Lakshman Ballal Joglekar and Lakshman Ballal Bhave, who were some of the witnesses from upper caste *Brahmin* families placed a brief statement of the past history of the *rayot's* forest rights so that on reading that statement commission may better understand the past. Manu, one of the wisest and ablest of Hindu law givers says, "From the people he (the king) must learn the theory of agriculture, commerce and other particular arts." Manu hereby evidently means that the people are the best judges on these points, and their convenience, their customs and their opinion, is therefore the criterion of deciding whether the administration of forests, agriculture or the like is good or bad. One single individual, let him be the very king is not justified in acting against the wishes and wants of the people in such matters...In conformity with these great fundamental principles of our gracious sovereign, the Queen Empress, when she assumed the governance of this country pledged the royal word to the effect that all our legal rights and customs will be scrupulously maintained. She in her August proclamation says "We know the feelings of attachment natives of India regard the lands inherited by them from their ancestors", and we desire to protect them "in all rights connected therewith subject to the equitable demands of the state and we will in generally framing and administering the law due regard be paid to the ancient rights, usages and customs of India."[73] This particular petition was made mainly for timber trees like teak on private lands.

---

71 See, *Forest Grievance Committee Report*, 1927; Also see, *Kesari*, 18 October 1892.

72 *Kesari*, 18 October 1892.

73 *Ibid*, Petition, pp. 25–26.

The stringent provisions of the rules made by the 1878 Forest Act affected not only the poor but also the upper class people from Hindu and non-Hindu communities residing in the rural areas. The upper class people were worried about their rights over forest lands. As in the letter above, they were trying to draw inferences from the past, from the Hindu *shastras* to prove that they had certain rights over forest lands. Similarly, the non-Hindu upper class people too were alarmed with the provisions of the act. They too voiced their opinions through memorials in front of the commission. Following is an extract from one of such memorials. It was submitted by Parsee proprietors like Mr. Nanabhoy Bairamji, Jeejeebhoy and others from the Salsette *taluka*.

> "...These rules provide to the people great hardship, that no timber of forest produce shall be removed without a pass from the officer and that in respect of every such pass there shall be payable such fee, as the Conservator of Forests with the previous sanction of the government shall prescribe. Prayers followed by prayers in profusion, wherein the general complaint was that many encroachments were made upon private lands and forests by the settlement and other Forest Officers. That the past system worked a great havoc, as would be hereafter detailed, and that the definitions in the Act were so vague, unlimited and comprehensive that even the very air we breathe might be verging upon the scope of the meaning attached to the term 'Forest Produce', and that the natural rights of man were probably interfered with in many uncalled for ways and his natural freedom of action was altogether stopped.
>
> It would seem to be the clear object and aim of the legislature in framing the provisions of Section 34 that in appropriating lands for forests, reserved or protected rights of private individuals and the government should be determined according to "justice, equity and conscience." However, our enquiries and personal experience have enabled us to say without fear of opposition, that in many cases (and these cases, we will prove by parole and

documentary evidence to the utmost of our power) the object of the legislature has been frustrated. An illiterate ignorant claimant who had not the prudence or foresight to preserve his title deeds, if he had any he would not have suffered because he could not produce any documentary evidence in support of the claim he would urge to a particular area proposed to be taken up in any *taluka* or village. We can cite several notices and summons issued by the settlement officers to claimants and others, calling upon them to produce documentary evidence if they had any, thereby clearly implying that oral evidence was of no value. Within our experience we venture to say we have not seen a tribunal, which could consider oral evidence as no evidence at all. As to documentary evidence in such cases, it is generally difficult to obtain for plain and palpable reasons. The history of our claims will show that there existed no necessity for a document in former times, no objection of any kind ever been raised to such rights and the *rayots* first occupancy having begun when there was none else than God to transfer possession to them, absolutely no title was even dreamt of. Should we attempt to find any such law on the point from which the officer may have borrowed his procedure of proof, we fail there too. So in the absence of any sound reasoning or sound law the claimant's right is made dependent on written evidence only. The burden of proof again is improperly cast on the wrong party. Mark again what opportunity the claimant has generally of placing his evidence before the officer! The officer as a matter of fact; uncontroverted and introvertible, move from village to village and *taluka* to *taluka* and the poor claimant has to move with him; that is one disadvantage. Owing to this uncertainty of the officer's stay again the claimant is not generally able to get advice or advisor to place his matter before the officer. Thus, half the chance of a successful issue is gone. Half that remains is generally washed out by the fact, that generally these Forest Officers are in the habit of looking at the question

> for enquiry with the spirit of a partisan rather than a strictly judicial matter.
>
> Thus slowly and surely, man after man and village after village was made to give up his or its rights in regular succession in favour of contemplated forests, and thereby brought upon himself or itself manifold inconveniences and hardships. In thus subserving the wants of the Forest Officer, the *rayot* brought upon himself a detrimental curtailment of his own rights and was left to the mercy of heaven alone for the fulfillment of his own requirements, and thereby became an unwilling instrument of his country's own ruin."[74]

---

[74] This testimony picturises aptly the confusion and difficulties faced by the people due to the number of formalities to be fulfilled to prove their claims for certain rights. The stringent provisions of the forest rules made with the Forest Act No.VII of 1878, caused great excitement and alarm among the people. Petitions after petitions were submitted both to the local and supreme Governments. This is one such petition. "Memorial to the President and members of the Forest Commission, Thana by the proprietors of freehold villages with forests attached there to in the *talukas* of Salsette, zilla Thana, Bombay Presidency", in *Bombay Forest Commission Report*, Vol. IV, p. 37; as aptly pointed out by Mark Poffenberger and Chatrapathi Singh, "Illiterate villagers were often unaware that a survey and demarcation process was in progress. Further most rural villagers, especially tribal communities, had little experience with legal procedures for filing cases in courts. While their rights had some protection and channels of legal recourse were available, it is hard to imagine remote tribal Indian communities of the late nineteenth century being able to avail themselves of those channels". See, Mark Poffenberger and Chatrapati Singh, "Communities and the State : Re-establishing the Balance in Indian Forest Policy", *Village Voices, Forest Choices : Joint Forest Management in India*, (ed.), Mark Poffenberger and Betsy McGean, Oxford University Press, Delhi, 1996, pp. 56–67; Also one legal scholar noted "Given the level of illiteracy and unsophistication of many tribal communities, the rules of notice, appeal and settlement have provided only the slimmest protection against the arbitrary extinction, dimunition or reallocation of rights", See, J. M. Lindsay, "Law and Community in the Management of India's State Forests", 29 July, 1993, p. 38 cited in *Ibid*.

In all 542 memorials were received from the Thana and Kolaba districts, besides 53 from other parts. To these were added 101 others. Hence, the total number of memorials received was 696.

The enquiry was confined to the districts of Thana and Kolaba as the alleged grievances of the inhabitants of the Thana and portions of Kolaba districts had been for sometime, prior to the appointment of the Commission, brought very prominently to the notice of the government and the people in general. Any special grievance in connection with the forest policy and administration which might be felt in other districts of the Bombay Presidency would, it was believed, fall under one or the other grounds of complaints brought forward by the inhabitants of the North Konkan.

Finally, the exhaustive survey of opinions – official and non-official, British and Indian – was completed in 1887. The Commission claimed, that it gave priority to the objective of long-run preservation of forests and also meeting the subsistence needs of the villagers. Commercial marketing of forest produce was to be given importance, they suggested, only when it did not impinge upon the two objectives mentioned above.[75] The Commission reaffirmed the system by which the Forest Department took over the village common or wastelands. However, it criticised the department for sudden imposition of restrictions on subsistence use of these lands by the villagers, as it had suddenly deprived them of their subsistence base. The Commission wanted this system to be extended to village wastelands patiently and fully acknowledging the villagers' minimal needs.

The Commission recommended that the forest rules and regulations should be simple and consistent enough for villagers to comprehend and also simple enough to administer so that friction between officials and villagers could be minimised.

One of the Indian members of the Commission, Rao Bahadur Krishnaji Laxman Nulkar, a close associate of the members of the Poona Sarvajanik Sabha attached his criticism to the final conclusions. Though

---

[75] For the recommendations of the Commission, see, *Bombay Commission Report, op.cit.*, pp. 68–72.

he supported the recommendations of the Commission expecting that they were likely to ease the burden on the people, he did not refrain from attacking the way the Forest Department worked. He vehemently criticised the department for indulging in massive confiscation of land rights since 1878, and for the way corruption had grown in the Forest Department. The sudden imposition of restrictions on the use of forests and forest produce, he pointed out, had reduced the people to a choice between starvation and poverty.[76] He thought the Commission's new proposals were the least that could be implemented to curb the forester's rapacity and avert the grave danger of defensive violence from the peasants. Finally, he gave credit for having brought forward peasants' grievances in front of the Commission to the larger landholders of Thana, as their effort had helped in avoiding a wave of desperate violence.[77]

Two years after the submission of the Bombay Forest Commission's report, the government came out with a resolution largely ignoring its recommendations. The government increased the powers and responsibilities of the Collectors and Commissioners in the sphere of forests. The provisions suggested by the Commission for the purpose of grazing land, practice of *rab* and collection of leaves etc., were completely ignored in the resolution. *Kesari* opposed the resolution sarcastically stating "...the nature of the resolution can be imagined or understood by the fact that 'The Times of India' (the English daily during that period) congratulated Lord Ray for his work after reading the resolution and commented that finally Lord Ray had acquired some practical knowledge/diplomacy about the political affairs."[78] This also showed the extent to which the forest issue had acquired significance in the political arena. Despite this, the Forest Department continued to be the cause of concern for the rural population in the following decade.

While reporting about the tribes of Thana and Kolaba the Commission had stated "The prohibitions contained in the rules against

---

[76] *Bombay Forest Commission* Report, Vol. IV, 1887, pp. 213–218.

[77] *Ibid*.

[78] *Kesari*, September, 1889.

moving and taking forest produce without a pass, has come to many of these persons in the light of a death warrant and has practically placed the population of the hills at the mercy of the forest subordinates, who have not scrupled to use their power in most cruel and oppressive manner. A large proportion of this vast population, numbering about a quarter million of souls in the *zila* is now reduced to the cruel dilemma of perishing by starvation or of appropriating wood and forest produce and being treated and punished as criminals."[79]

In addition to the prevailing miserable condition of the rural population, the privilege granted to the tribals for collecting forest produce, especially firewood, for sale was withdrawn in 1894. This naturally resulted in violent confrontation with the government. The discontent was widespread. The tribals refused to work for the Forest Department. They boycotted the contractors. They violated government orders by cutting fuel wood from reserved forests. There were attacks on petty government officials. The office of the Deputy Collector was destroyed. This protest continued until the resolution was withdrawn.[80]

Instances of breach of forest laws, organisation of public meetings to demand the lowering of land assessment, withdrawal of restrictions pertaining to the forest etc. occurred frequently. The media continued to highlight the problems of the people concerning forest issues. The torturous rules and regulations of the Forest Department and their implementation by the most ruthless forest officials continued to be a subject of criticism. In the 23 January 1906 issue, *Kesari* compared the Forest Department with a devil who had made peoples' lives miserable. It appealed to the people to organise themselves and protest collectively against the new unjust resolutions.

---

79 *Bombay Forest Commission Report*, 1887, Vol. IV, Vol. 4, p. 42.

80 This protest also gave vent to other grievances such as *toddy*, grazing and *abkari* regulations. People demanded that the palm tree tax be abolished. They wanted that country liquor as well as salt and husked rice be sold cheaper and that the government redeem the mortgaged lands of the *rayots* from the *savkars* and restore them to their rightful owners; Indra Munshi Saldhana, *From Shifting Cultivators to Peasants and Labourers : A Historical View*, 1992, pp. 1–19.

According to the Administrative Report of 1908–1909, the Forest Department of the Bombay Presidency declared a net profit of Rs 14,94,834. The number of cattle impounded during that year in the Bombay Presidency was as high as 1,90,605 out of which 1,35,761, i.e. 71 per cent belonged to the Central Maharashtra, where forests were very less unlike the Konkan districts or *Khandesh*.[81]

Central Maharashtra always suffered from shortage of rainfall, and hence was devoid of good forests. However, with rich grasslands available, it formed the main pastoral region of Maharashtra. But major parts of these grasslands got grabbed by the British Forest Department, resulting in shortage of grassland for the *rayot's* cattle. Consequently, the extent of impounding of cattle in this region was very large.

To avoid this situation, some amount of free grazing land was demanded by the villagers. Also, the grasslands where villagers could graze their cattle by paying fees were sought to be under the responsibility of the village panchayats. Though the Mumbai Government had even passed a resolution recognising that "Forest Department was one of the departments which had to work in close association with the villagers. Hence, people's participation in its administration was highly desirable..."[82] the formation of village panchayats took a back seat. The Forest Department did not want to displease those who were collecting the forest fees until before the resolution.[83] It was feared that those who collected the forest fees till then were the elite of the villages, and hence they were not likely to allow the establishment of village panchayats. Hence, in spite of the support from the media, this resolution which could have worked in favour of the common villagers was given up by the government.[84]

---

81 *Kesari*, 14 July 1910.

82 *Kesari*, 12 July 1910; However, many British Officers were not happy with this decision. They were suspecting involvement of political interests in the villagers' demand see, *Forest-AntiForest policy of the Bombay Government*, Government of Bombay, 1899.

83 *Ibid*.

84 *Ibid*.

However, some positive steps were again taken by the Mumbai Government in October 1910 to improve the situation of the villagers in Central Maharashtra, also in the upland area, thanks to the positive recommendations of a committee that was appointed to study the reasons why the number of impounding cases was extremely high in upland Maharashtra. The committee criticised the old policy of the government in the uplands. Though late, the government finally accepted the recommendations of the committee such as no grazing lands, if they had to be closed, should be closed for grazing for more than 10 years. No forest land, which was within less than 5 miles distance from the village, was to be prohibited for the use of the people. The village in which people helped to extinguish fire was to be exempted from the payment of grazing fees etc. Also, with these recommendations, the forests of grassland were made the responsibility of the revenue or civil department called *Mulki* Department in Hindi. The Solapur district, Akola, Sagamner, Kopargao, Srigonde, Nagar district, parts of Satara district, Nijad *taluka* of Nasik district and Chalisgao of East Khandesh were the regions where the forests were transferred and left under the charge of the revenue or civil department.[85]

This decision of the government was welcomed by the people.[86] However, besides a few minor changes as mentioned above, the forest policies more or less remained the same and continued to be a matter of concern for the rural population of the presidency. The Forest Department becoming a transfer subject did not help improve the situation, except for the fact that more and more people started coming forward with their grievances. As a result, Forest Grievances Committees were again formed in 1924 and then in 1935, though without yielding any substantial positive results.

The Forest Grievances Committee appointed by the Revenue Department Resolution No. L.C.1374 dated 26 January 1924 was inclusive of official members as well as non-official members. It

---

85 *Kesari*, 25 October 1910.

86 *Ibid*.

included the Chief Conservator of Forests, Bombay Presidency, three Conservators of the three circles and five Indian members of the Legislative Council, Bombay. This Committee received a large number of petitions of grievances from the rural population, some of them common to nearly all the districts of the presidency, some peculiar only to a few districts and some pertaining to single villages or even individuals. The people of 11 out of the 13 districts from which petitions were received were interviewed during the tour, and matters were enquired into on the spot. Their complaints were mainly regarding the harassment by the subordinate forest staff, hike in the grazing fees, insufficient availability of grazing areas, impounding of cattle in unjust manner, unjust or communal punishment in case of forest fire and so on.

However, the most interesting feature of this Committee was the wide differences between the view of the official and non-official members on many very important points. The differences were so sharp that on some points it was impossible for them to write a unanimous report. Finally the Committee came forward with two separate reports; one representing the views of the official members and the other those of the non-official members.

The views expressed by these two separate reports only express the different attitudes with which the problem of forestry was approached. The differences were very contrasting. The opinion adopted by the officials was clearly very rigid and unsympathetic towards the people and was meant to give first priority to the commercial needs and enhancement of revenue for the government. For example, one of the most important grievance of the people was that there had been a rise in the grazing fees from 2 *annas* to 4 *annas* since 1922. To this the official members answered that there had been a rise in the price of all other commodities, including agriculture produce, which was presumed to have put the agriculturists in a position to pay a higher fee than the purely nominal one or two *annas* formerly charged. The Committee was obviously, not in favour of the abolition of fees as it brought a very considerable revenue to the government – Rs 3,53,655 for village cattle in 1923–1924 and Rs 3,83,989 in 1924–1925.

It was pointed out to the Committee that the agriculturist paid land revenue for the land he cultivated. It was proper to charge him a nominal fee, a low one at that, for the valuable privilege of grazing on government land as the cattle were those required for cultivating the land, on which he paid land revenue. But the official argument was that the government intended to charge the nominal grazing fee only to prevent the establishment of a claim to the right of free grazing. The official members were also of the opinion that 4 *annas* fees in most cases was much below the real value of the grazing enjoyed. Hence, it could justifiably be considered a nominal fee.[87]

The unsympathetic attitude of the official members of the Committee was obvious from their answer to the people's complaint against the harassment by lower class forest officials. The official members chose to believe that the people's complaints were grossly exaggerated. Their stand was that the forest guards were after all local men known to their neighbours. They stated "It is curious that the mere fact of a man being appointed as forest guard should at once change him from down-trodden uncomplaining serf to an arrogant oppressor...On the other hand, cases of assault and even of murder of forest guards are not at all uncommon."[88]

They even refused to appoint committees of locally influential people to supervise the work of the forest staff. The committees, they thought, could unnecessarily lead to trivial complaints against the forest staff in order to stop them from doing their duty.[89]

In contrast to this, the non-official members were very sympathetic towards the people's problems. They boldly stated in their report "As we differ on certain important points from the official members of the Committee and as we have to bring to notice certain other important points, it has become necessary for us to submit our separate report." The non-official members sympathised with the

---

87 *Forest Grievance Enquiry Committee*, Mumbai, 1927, Part I – A, pp. 25–30, Part II – B, pp. 21–27.

88 *Ibid.*

89 *Ibid*, and Richard Tucker, "Forest Management and Imperial Politics." *IESHR*, pp. 298–299.

complaint of universal harassment of the *rayots* by subordinates in the Forest Department service. In fact, through several applications, they tried to bring to the notice of the Committee the gravity of the problem. For instance, one of the applicants said "Any man with sense of self respect will feel so much annoyed at the forest grievances that he would come to think, that these grievances would never be redressed without violence. Sir, please sit for 5 minutes at the place where passes for grass and fuel are issued on receiving payment. You should be a real saint if the conduct of the subordinates would not fill you with disgust. It will take years of enquiry and decision in this matter. No improvement can be expected from the Committee. The more we think about the matter, the more despondent we feel."[90]

The non-official Committee showed the willingness to notice and understand people's skepticism about government committees, based on their past experiences and the resulting fiascoes. The non-official members stated "The second thing we prominently noticed, is the people's belief that no redress would ever be given to them. They hold that there is no forest grievance of the people which is not known to government or at least to the superior officers of the government in forest service. They argue that all their grievances are too well known to the government and if the government are really sincere, they could have removed these grievances without having appointed a committee."[91]

About the grazing question, they went into the history of grazing fees. They stated that in spite of repeated cancellation of rise in grazing fees, it was once again imposed in 1921. The non-official members stated "We know our people better. They groan but submit, they do not complain unless someone is found to take their cause. There was murmuring but no loud noise. Emboldened by the helpless submission of the people government raised it (the grazing fee) in few more districts in 1922–1923 and everywhere in 1923–1924. Income raised as follows:

---

90 *Forest Grievances Inquiry Committee Report*, Part I – B, Governent Press, Bombay, 1926–27, p. 24.

91 *Ibid*.

Income from grazing in 1920–1921 was Rs 2,79,846, Rs 3,13,408 in 1921–1922, Rs 4,85,804 in 1922–1923 and Rs 5,21,835 during 1923–1924."[92]

Finally, there was an agitation. Applications were made to government, questions were asked and resolutions tabled in the council. The present committee was appointed under the circumstances mentioned.[93]

In the end, the Committee reminded the government of its policy laid down in 1891 "In its management of forests the ease and contentment of the people is an object of greater solicitude to government than the realisation of revenue, and while no relaxation of precautions necessary for the conservation and reproduction of the timber and firewood supplies can be permitted, and the Forest Officers must be vigorously supported in resisting unauthorised encroachments, government has no desire to increase their forest revenue by curtailment of conceded privileges or of local supply, or by the levy of excessive charges for grass and other minor forest produce. The benefit of any revenue so obtained would be altogether insufficient to countervail the hardship and irritation that would be caused, were the *rayots* unduly pressed in the matter of obtaining *rab* and firewood, grazing for their cattle or grass for thatching and other household purposes."[94]

The non-official members of the Committee urged the government to stop the increase in grazing fees and fees on other minor forest produce without a moment's delay. They warned the government of the discontent and irritation that was caused by the increase in the fees. They also showed their disagreement with the other (official) committee members by stating, that though the prices of agricultural goods had gone up, the profits made due to that were

---

92 See, the *Forest Administration Reports*, 1921–22, 22–23, 23–24, 24–25.

93 *Ibid*, A reflection of the British attitude towards the issue of grazing can be discerned from the article "Forest Pastures in the Deccan", R. A. Fagan, *Indian Forester*, Vol. 11, April 1885.

94 *Ibid*, vide Bombay Forest Manual, Vol. II, pp. 217 and 312.

going away in the increased cost of agricultural production. Cost of articles required for cultivation had also gone up.

About the imposition of communal punishment, for example in case of forest fires, they urged the government to do away with the practice. They brought to the notice of the government the fact that "People had come to look upon the vicinity of forests to their villages as a curse rather than a blessing. They suffer more on account of that vicinity than gain by it."[95] Hence they recommended the system of giving rewards due to which the people themselves would stand to gain or lose by the fires. Finally, they reminded the government of its own policy that "the sole object with which State forests are administered is the public benefit,"[95] and if it was adhered to in the true sense, it would make the people of the Presidency really happy.

The glaring contrast in the attitudes seen in the two reports of the same committee was probably also a reflection of the political situation in the first half of the twentieth century. By the second decade of this century the political struggle challenging British authority had assumed a much more explicit and aggressive form. The politics of mere submission of petitions and memoranda was left behind. The introduction of 'Diarchy' created a new sentiment among the people. The Indian nationalists were no longer willing to let an alien government decide their fitness for self-government, nor were they satisfied with the shadow of political power. Masses were being organised under the leadership of Mahatma Gandhi and they had begun to fight for their own rights. Expression of this was found in case of Bombay Presidency in the form of *Jungle Satyagraha*. The year 1930 witnessed such *satyagrahas* in almost every nook and corner of the Bombay Presidency.[96] The forest issue was no more an issue to be solved by way of requesting and petitioning the government. Through

---

95 *Ibid*, Part I – B, p. 31.

96 At one of such *Jungle Satyagrahas* near Panvel at Chirnet, police opened fire. Though there were teak trees in the forests, the *Satyagrahis* went and cut only firewood and were returning peacefully when police opened fire and people were arrested. *Kesari*, 30 September 1930.

Jungle Satyagrahas people, began to assert their rights over forests instead of pleading the government, as done in the past. In fact, often the forest issue became a tool for the Indian nationalists to reach out to the masses in the Indian countryside and this issue even brought the tribals from the remotest parts of the Presidency into mainstream politics. According to the reports of the Marathi newspaper *Kesari*, the *Jungle Satyagrahas* organised during 1930 everywhere in the countryside, brought together people from various sections of society, the elites like doctors and advocates, tribals like *Bhils*, *Kolis* and poor peasants, the most backward classes like *Mahars* and *Mangs* and the forwards like the *Chitpavan Brahmins*. For example, during the *Satyagraha* which took place in western *Khandesh* on the day of *Gokulashtami*. First, the *Satyagrahis* performed puja to photographs of Krishna and Gandhi and then they cut grass and firewood to symbolise the breaking of rules.[97]

In fact, *Jungle Satyagraha* became a popular tool, especially in the hands of the tribals, predominantly from forest regions of Bombay Presidency. The first *Satyagraha* in Nasik district was at Bhilwad. However, the *Satyagraha* at Saptashringi was the biggest in Nasik district. Piraji Sayaji Gavit of Devalgaon, Kalusoma Kunwar of Palsadargaon, Davalbagoo Laware of Deshgaon, Sakharam Thakre of Chinchapada and Damu Waghmare of Aamchas were the young tribals who created awareness amongst fellow tribals about their rights over forests and made the tool of *satyagraha* popular in the jungles of Western Ghats in the Bombay Presidency.

In Nasik district, as mentioned, the young tribals actively created awareness about *satyagraha* in the countryside during the months of July, August and September of 1930. On 18 October 1930, people were collected together by beating drums and screaming by the volunteers at Chanakpur in Kalwan *taluka*. On this occasion, around 10,000 – 12,000 tribals gathered together with their weapons like bows, arrows, axes and sticks etc. though *satyagraha* was never supposed to be performed by violent means. In order to suppress this *satyagraha*, the police and

---

[97] *Kesari*, 23 August 1930.

the army surrounded the hill. The *adivasis* there had only one aim – to put Gandhi Baba on the throne.[98] The British officers tried to negotiate with the leaders of the *satyagraha*, though in vain. The whole of 18 October night was spent by the *satyagrahis* on the same hill surrounded by the police and army. Next day, police opened fire to disperse the people. As the police opened fire, the *adivasis* retaliated with bows and arrows. Seeing this, the police increased their firing without discretion. This *satyagraha* which went on for one and a half days witnessed at least 100 deaths and many were injured. Six leaders of this *satyagraha* were punished by the government.[99]

Without being discouraged, *satyagrahas* continued all over the Bombay Presidency, of which, the one at Junnar in Poona district, Akola in Nagar district, and one at Ambegaon went down in history. Hundreds of people were imprisoned and fined during these *satyagrahas*.[100] As a result, according to the 1935 Act, certain regions of Bombay Presidency, such as Panchamahal, East and West Khandesh and Nasik and Thana district were declared as *Amshataha Wagalalele Kshetra* (partially excluded areas in Bombay Presidency) and the Bombay Governor was made directly responsible for this region.[101]

However, this did not deter people from asserting their rights over their own land as by now the struggle for independence had

---

[98] Similar protests against forest regulations were made by the rural population in Andhra Pradesh, since the early decades of the twentieth century. However, non-cooperation movement as envisaged by Gandhi did not give prominent place to these protests. See, Atluri Murali, "Civil Disobedience Movement in Andhra 1920–22 : The Nature of Peasant Protest and the Methods of Congress Political Mobilisation", *Congress and Classes: Nationalism, Workers and Peasants*, (ed.), Kapil Kumar, Manohar Publications, New Delhi, 1988, pp. 153–175.

[99] Dr. Govind Gare, *Sahyadriteel Adivasi Mahadev Koli*, (Marathi), Sahadhyayan Publishers, 1974, pp. 85–90.

[100] *Kesari*, 19 August 1930 and 23 August 1930.

[101] For more details regarding this partially excluded areas, see D. Symington, *Report on the Aboriginal and Hill Tribes of the Partially Excluded Areas in the Bombay Presidency*, Bombay, 1939.

gathered momentum. Along with Indian National Congress, there were several other organisations springing up all over the country. Some of these were inspired by the Russian Bolshevik Revolution of 1917. During the 1930s, B. R. Ambedkar's Independent Labour Party, the Kisan Sabha, whose leaders were Indulal Yagnik, Dinkar Mehta, Kamla Shankar Pandya, along with the Indian National Congress were engaged in mobilising the agricultural tenants and tribals in the Bombay Presidency. These organisations were not satisfied with the measures being adopted by the Congress Ministry for the welfare of the rural agricultural population.[102] In their effort to extend their base in the countryside, the different political parties were competing with each other.

The *Varli* Revolt of 1945 in the Thana district was one example where decades of exploitation of the *Varli* tribals was given voice.[103] The entry of the Kisan Sabha in Thana in 1945 provided the necessary ideological and organisational support to the *Varlis* against the oppression of landlords. As mentioned already, the plight of the tribals in the Bombay countryside was pathetic. The landlord tenant relationship had assumed its worst possible forms. The rentals were too high.[104] To top it all, the stringent forest rules did not allow them any alternate source of income. The landlords in their exploitation of the *Varlis* were assisted and supported by government officials and their own private staff. Without the support of government officials this exploitation would not have been possible. However, Forest Officers, circle inspectors, *Talatis*, police inspectors were obliged to the *savkars*. Whenever any of these government officials were on their

---

[102] *Peasant Mobilisation, Political Organisations and Modes of Interaction, The Bombay Countryside*, 1934–41.

[103] Indra Munshi Saldhana, *From Shifting Cultivation to Peasant and Labourers: A Historical View*, pp. 1–19.

[104] D. Symington also reported the prevalence of forced labour *veth* demanded by the landlords from their tenants. For details see, D. Symingon, *Report on the Aboriginal and Hill Tribes of the Partially Excluded Areas in the Bombay Presidency*, Bombay, 1939.

official visit to the village, they were the guests of the *savkar* or the zamindar of the village. They stayed in their palatial houses and were given lavish treatment. They travelled in the vehicles of these zamindars. If the officer was of a higher cadre, the zamindar himself travelled along with him and looked after his guest. For lower level staff, zamindars' personal servants were given the responsibility of hosting. It was only after the tribal movement of 1945 that the tribals started seeing the forest officials camping in tents. As soon as a new government officer or a Forest Officer was appointed, there was competition among the zamindars and the forest timber contractors to bring them into their own fold. Regular supply of milk, *ghee*, grains, fish and meat to these officers' residence was also a common practice. The tribals were witness to all this. They were not even aware of the fact, that the government officers were supposed to perform their duties impartially, for the benefit of all and not only for the zamindars or forest contractors. In fact, the zamindars while physically torturing the tribals used to say openly, "who is going to sympathise with you all. All the police, *Talatis*, *Mamlatdars* and also the judiciary are on our side. There is no one who can help you out. Even if we take your lives, there is no one to even question us."[105]

Forest Department staff was literally their enemy. Leaning, sowing, planting in the forests, extinguishing the forest fires were the duties of the tribals. However, the timing of the tree plantation programme in forests and their personal farming always coincided with each other. Hence, they had to keep aside their personal work in order to fulfill their duty towards the Forest Department. If in some way they tried to avoid the work in the forests, Forest Officers dragged them to the forest and made them work without any payment in return. The wages which they were supposed to receive never reached them. They were taken away by forest staff before they could reach them. For them working for the Forest Department was the same as working for the zamindars, that is as bonded labourers. Once one tribal told

---

[105] Godavari Parulekar, *Jnhva Manoos Jaga Hoto*, Mouj Publishers, Bombay, 1969, p. 48.

one Communist activist, Godavari Parulekar,[106] with amazement that "Once one Forest Officer actually gave us salt for having worked in the forests!" They were not even aware that they were supposed to be paid by the Forest Department for their work. If the work in the Forest Department was denied by the tribals, false charges were made against them and they were punished. They were never even addressed properly by the officers. It was only abusive and dirty vulgar language that was used for these poor tribals. Later with gradual awareness among the tribals, some amount was paid to them, though it was measly.[107] An example of this hostile attitude of the Forest Officers towards the wild tribes is a letter, where the correspondent officer called himself a 'victor' expressing the view that the forest tribes were an anachronism and ought to be deported with bag and baggage from their homes, while he proposed to treat the young *Bhils* and *Varlis* as 'panther cubs' confining them in iron cages with a chain around their legs.[108]

In the sale of grass, the contractors from the city made thousands of rupees every year and became *crorepatis,* but paid only insignificant and measly amounts to the tribals. During the grass cutting season, hundreds of tribals were employed in the Thana countryside. Whether it rained or it was cold they had to perform their duties. When it was cold after the day's hard work they would light fire and lie down besides it at night in the forests. For all this hard work, instead of getting good wages all they got was some 3 to 4 *annas* and a whip once in a while, which was given by the supervisor in order to create fear in the minds of these labourers so that they worked properly. They had to work relentlessly. They could not even stop to drink water or smoke a

---

[106] She was a Communist activist mainly instrumental in the uprising of *Varli* tribals against bonded labour. She entered the political scene in the Thana district in 1945.

[107] Godavari Parulekar, *Jnhva Manoos Jaga Hoto,* Mouj Publishers, Bombay, 1969, p. 49.

[108] A letter to the Editor of Bombay Gazettee, "Forest Conservancy in Bombay", *Indian Forester*, September 1886, p. 305.

beedi. At the time of World War II, the grass contractors made crores of rupees. But there was no rise in the wages of these tribals. Similarly, in 1946 when timber for construction and firewood was in great demand, the timber merchants and contractors made crores of rupees. But the tribals, however hard they may have worked, were paid only 4 *annas*. Such was the state of the tribals.[109]

Finally, such amassing misery erupted into a revolt directed against the moneylenders, landlords, traders, Forest Officers, forest contractors and their musclemen. The tribals put an end to the age old practice of *Veth bigar*. 'Get free from slavery' was the call given to the tribals and hundreds of them were liberated.[110] The tribals succeeded in regularising the rent they were obliged to pay for the land they tilled. The dispute between the tribals and the landlords, as well as the timber merchants regarding the grass cutting and felling of trees in the forests, was settled in favour of the tribals. Tribal women also participated actively in the struggle to do away with the oppressive and humiliating practices of the landlords, moneylenders and forest guards.

Repressive measures were taken by the State to end this struggle. Large numbers of activists were arrested; prominent workers of the Kisan Sabha were externed from the Thana district. Several times police resorted to firing. One tribal organisation, though it was run with the help of the government and the Congress Party, declared in *Bombay Chronicle* of 25 October 1945 that "whether the firing was justified or whether it was excessive, we were not able to ascertain. It is, however, clear that the police were firing from the roof of the van which was converted into a sort of armoured car. No one from the police seems to have been hit, though there was stone-throwing. We feel that there should be an open and independent enquiry into it."[111]

---

[109] Godavari Parulekar, *Jnhva Manoos Jaga Hoto*, Mouj Publishers, Bombay, 1969*t*, pp. 49–90.

[110] *Ibid*, S.V. Parulekar, *Revolt of the Varlis*, Popular Publishing House, Bombay, 1947.

[111] *Ibid*, p. 85.

A state of emergency was declared in the Thana district. Finally, this phase of the struggle more or less came to an end in 1947. The tribals succeded in regularising the rent they were obliged to pay for the land they tilled. The Maharashtra Congress Committee appointed a sub-committee to enquire into the cause of the struggle. It brought out a report recommending immediate reforms in the system of land tenure in the region.[112]

To conclude, establishment of the Forest Department and its policies had a far reaching and unprecedented impact on the life of the rural populace of the Bombay Presidency. An agrarian nature being the basic characteristic of this society during the Maratha period, the lives of people were closely dependent on the forests, though in varying degrees. Commercialisation in every sphere of the economy resulting into exploitation of natural resources for commercial purposes, combined with new technology, only intensified its negative impact.

Depending on the circumstances, severity of the impact and the intellectual input, different sections of society reacted to this impact in different ways. The spontaneous uprising of tribals in the initial period was a response to the new system and to the new beneficiaries of the system, like the moneylenders or the landlords and the government officials. This was meant to overthrow this class of elites, who were a creation of the new system.Though it did not have much impact at national level politics, it definitely succeeded in bringing about some kind of awareness regarding their problems in the minds of the forest officials.

Forest offences were another form of unorganised protest made at the individual level, which continued to worry forest officials throughout their rule. Forest fires were one of the most suspected offence.

Legal protests made by way of petitions were aided by the national leaders. Such protests involved people from all walks of life.

---

[112] Indira Saldhana, *From Shifting Cultivators to Peasants ad Labours: A Historical View*, pp. 16–19; Also see Godavari Parulekar, *Jnhva Manoos Jaga Hoto*, Mouj Publishers, Bombay, 1969, pp. 115–118.

However, the testimonies belonging to the poorer strata of society show that they were more concerned about needs at a subsistence level of living such as lopping of trees for rice cultivation, prohibition on shifting cultivation, fuelwood etc. Their protests were aimed at the petty forest officials, namely the forest guards, who harrassed them and made their lives miserable. Whereas the upper strata of society like the landlords and moneylenders were concerned about the restrictions on the availability of timber. Nevertheless, all sections of society were discontented due to the alienation of the age old rights over forests.

Institutions like The Poona Sarvajanik Sabha, the members of which belong to the forward castes of Mahrashtra, provided strong support to the rural population in these protests in order to represent their grievances. Several intellectuals of the period were concerned about this issue. Though in principle they agreed about the importance of forest conservation, the way in which it was implemented was not agreeable to them. Hence, they gave voice to people's grievances in several ways such as newspapers or petitions. In the later period, the issue of forestry was also used by different political parties as an important tool against the foreign authorities. For example, Congress used it by way of *Jungle Satyagraha*, though these *Jungle Satyagrahas* were not always restricted to Gandhian methods such as non-violence.

Environmental crisis has become a matter of major concern during the last few decades. Studying the history of people's resistance to resource management policies thrust from above by governments in the past may be of use in formulating policies in the future. By this, it is possible to avoid the mistakes made by our predecessors and move away from Eurocentric models of development, and consider the possibility of communal control over natural resources combined with a sustainable exploitation pattern.This is essential as the people's acceptance of government policies is crucial for them to become successful.

8

Conclusion

The formation of the Forest Department, imposition of several Forest Acts and their application by way of day-to-day administration have had an unprecedented and everlasting impact on Indian society.

From the earliest periods of known Indian history, there is a record of human interaction with and human intervention in natural surroundings. However, the major difference between the pre-British period and British period lies in the scale and magnitude of this intervention, and the everlasting impact it has had on the society. An attempt has been made in this book to study the formation of the Forest Department, the evolution of its policies and the unprecedented impact it has had on the society in Maharashtra.

The basic assumption underlying the formation of the Forest Department by the British was that Indians lacked the sense of conservation, and hence there was a need to introduce scientific conservation of forests. However, the more compelling factor was fulfillment of the timber requirements of their home country. In order to fulfill these demands, there was a need to legitimise the exploitation of forests which was attempted by way of passing several Forest Acts.

The geography of the region, its climatic conditions and the types of forests in the Bombay Presidency along with their commercial potentialities, had a definite bearing on the evolution of British forest policy.

The history of Maharashtra along with the socio-economic conditions prevailing in this region during the medieval period reveals that there was a close inter-relationship between political interference and the level to which natural resources were exploited. In the pre-British period there was a definite emphasis on the management of natural resources, particularly the forests in Maratha region (i.e. Bombay Presidency). Despite the absence of an independent Forest Department, there were certain ways in which attempts were made by the Maratha rulers for the preservation of forests, although to a very limited extent. The decentralised power structure gave considerable autonomy to the villages. To that extent the control over forests was also much less as compared to that of the British period. However, it was not only the rulers who had assumed this responsibility. By way of its customs and traditions, society as a whole helped the process of conservation of natural resources.

The first crucial interlude had been the structure of early British management of forests, which could literally be called management of teak plantations. This is because the main concern of the company officers was fulfillment of imperial timber needs. Several changes took place in the British forest policies from time to time depending on the demand for timber and need to increase land revenue. However, what is striking in this period is the different opinions expressed by different officers regarding the management of forests, which reflected their understanding of the society or the lack of it. For instance, a statesman like Elphinstone, who was influenced by the thinking of Thomas Munro, believed in minimum interference with the old customs and traditions of Indian society.[1] He opted for a policy of conciliating and pacifying the socially dominant classes of Maharashtra without alienating the peasant. For example, his proclamation assured the land holder that their tenures would be preserved.[2]

---

1 See Chapter 3 of this book.

2 Consequently the British retained the privileges of *Jagirdars*, whose loyal attachment was much sought after during the crtitical phase, when the East India Company was having a precarious hold over the newly-conquered people. See, Sushma Varma, *Mount Stuart Elphinstone in Maharashtra : A Study of the Territories Conquered from the Peshwas*, Bagchi and Company, Calcutta, New Delhi, 1981, p. 259.

Also, there were Collectors like Dunlop, the Collector of Ratnagiri, who believed in imposing no restrictions on the peasants as far as forests were concerned.[3] These officers were often at odds with more overzealous officers, who were keen to exercise their seignorial rights over forests for the fulfillment of imperial demands for timber and to increase revenue.[4]

These differences of opinion culminated into a prolonged debate on the eve of the passing of the Forest Act of 1878. The extent to which privileges should be granted to the people in the forests was a matter of debate among the Forest Officers. The so-called scientific and technological backwardness of the rural people was held as enough justification for the State control of forests by the annexationists. The passing of the Forest Acts symbolised political power to the forest staff in the realm of forest management and the assertion of sovereign rights over the colonised land.

The populists and the pragmatists showed some understanding of the people's customs and the sustainable practices as far as agriculture and forest operations were concerned.

Inter-departmental rivalry between the Revenue and Forest Departments turned out to be a hindrance to the increasing control of the Forest Department over the forests. Being compelled to generate and enhance revenue, officials of both the departments were under pressure to transform land into a revenue generating source. The revenue officials thought that they were better equipped to protect trees, because of their familiarity with the rural population and their needs. Whereas the overzealous Forest Officers lay their emphasis more upon efficiency in forestry at the cost of all other interests. However, all officers favoured imperial intrusion in land management to protect and improve timber stocks, though they did not agree on the extent of privileges to be granted to the people and the extent of regulation to be imposed.

---

3 R. D. Choksey, *History of Bombay - Konkan 1818–1938*, Bombay, 1957.

4 See, the opinion of Underwood in W. R. Baillie, *Summary of Papers Relating to the Madras and Bombay Forests*, Bengal Secretariat, Military Orphan Press, 1851, p. 204.

These ideological underpinnings which influenced the colonial forest policies had a crucial bearing on the human-forest interface. Hence, an enquiry has been made into it in the present book. Also, the Forest Acts of 1865 and 1878 and the Forest Policy of 1894 have been discussed in detail. The 1865 Act empowered several local governments to declare certain areas as State forests without in any way interfering with the rights of the people. From the point of view of administrators, this Act was imperfect and unsound in principle. The administrators criticised the law for being made and enforced locally without paying attention to their legality. They found the claims for forest rights so overwhelming that real forest conservancy became practically impossible. Hence the 1878 Forest Act was passed. With this Act the reserved forests secured a permanent settlement, whereas in protected forests existing rights were recorded but not settled. Originally it was thought that a limited protection would be sufficient to the lands which could not at once be decided as suitable to form reserved forests. Soon most of the protected forests were converted into reserved forests. The thinking behind forming village forests was that village communities and private persons had a right to expect that the government would provide protection to their forest property. The village forests were to be under the management of government officers.

That the shaping up of the forest policies depended on the needs of the time, has also been examined here. For example, to fulfill the demand for timber during the two World Wars, several changes were made in the forest policies.

The natural outcome of the evolution of these forest policies was alienation of the local population from their traditional means of subsistence and production systems. Increasing commercialisation of forests for timber production and other forest produce, replacement of species useful for the tribals with the commercially viable species like teak, along with several restrictions on the use of the forests made matters worse. The branding of the rural population as basically wasteful, destructive, short-sighted and inefficient combined with the assumption of their so-called technological and scientific backwardness gave the British administration moral justification to undertake this process of alienation successfully and rapidly. Behind all this was the inherent dislike among the British administrators for the people on the fringes of the society.[5]

Alienation from their traditional means of livelihood and the sudden imposition of a sedentary lifestyle caused endless misery and confusion in the minds of mobile forest dwelling communities. This erupted into several forms of protest like individual offences, communal protests and revolts.

Legal protests made by these people were in the form of petitions to the government, in which they were aided by contemporary national leaders like Tilak and Ranade. Institutions like the Poona Sarvajanik Sabha, the members of which belonged to the forward castes of Maharashtra, provided strong support to the rural population in these protests in order to represent their grievances. Several intellectuals of that period were concerned about this issue. Though in principle they agreed about the importance of forest conservation, the way in which it was implemented was not agreeable to them. Hence they gave voice to people's grievances in several ways.

Several revolts which took place during this period were many a times not important from the point of view of national politics. However, they succeeded in bringing about some kind of awareness regarding people's problems in the minds of the forest officials.

The present study does not assume that there was a perfect ecological equilibrium in pre-British India. In fact, historians like Romila Thapar have warned against any kind of romanticisation of resource management in the pre-British India. For instance, she points out that even during the Mauryan period, from the fourth to second century BC, the State endeavoured to enhance its revenue by clearing forests to extend agriculture. State initiatives to extend arable land and establish settlements thus took a heavy toll on Indian forests. However, this was kept under control to some extent, as cutting of trees without

---

5 See, Mahesh Rangarajan, *Fencing the Forests: Conservation and Ecological Change in India's Central Provinces 1860–1914*, OUP, Delhi, 1996, pp. 198–201.

permission from the State was prohibited.[6] One of the possible reasons for the decay of Indus Civilisation has been attributed to environmental degradation.

Attempts to extend agriculture by revenue farming or by providing direct incentives by way of tax redemption were not at all uncommon during the medieval and late medieval period in the Bombay Presidency. In fact, as a response to financial crisis farming of the revenue had become a general practice after 1761.[7] Peshwa Madhav Rao I wrote in his will in 1772 that both in Konkan and *Desh* the *rayots* were oppressed due to revenue farming and he pledged for its abolition and substitution by regular management.[8]

However, the scale and magnitude of these operations were much less by virtue of the level of technology and the level of demand as compared to that of the British period. Sophisticated means of deforestation like modern machinery, in combination with modern means of transport and communication, helped the British Forest Department to undertake unprecedented forest utilisation operations much more efficiently along with economisation of time. As the economist C. H. Hanumantha Rao puts it "So long as land is plentiful, nature bountiful in relation to population and demand, management and technology could be ignored without much loss at least in the short run."[9]

With the advent of the British the demand for timber, especially teak, increased several fold. This demand was not only due to the requirements in India, but also the requirements of their mother country.[10]

---

6 Namita Roychoudhary, 'Past lessons and future strategies,' *Down to Earth*, 31 July 1992, pp. 33–36.

7 Revenue farming is a practice in which the rights to collect tax are sold to the highest bidder. Andre Wink, *Land and Sovereignty in India*, published in association with Orient Longman, CUP, 1986, pp. 321–324.

8 *Ibid*, p. 341.

9 C. H. Hanumantha Rao, *Agricultural Growth, Rural Poverty and Environmental Degradation in India*, OUP, 1994, p. 165.

10 This valid point was emphasised by the newspaper *Kesari*, 9 June 1885.

Initially, there was indiscriminate exploitation of forests to fulfill the demands of the Royal Navy. The Royal Navy was instrumental in satiating the British imperial interest in expansion all over the world. However, later it was realised that Indian forests were not inexhaustible. Hence, attempts were made to preserve teak. In fact, one can state that initial conservation of forests was only teak conservation. This sharply contrasts with the claim that is made by some writers that British were the first ones to introduce scientific forestry in India.[11] No doubt there were a few Forest Officers like Dr Gibson and Hugh Cleghorn in the early phase of the establishment of the Forest Department, who pointed out the connection between deforestation and drought, the need to protect forests to maintain water supplies and safeguard agricultural prosperity. They claimed that the destruction of forests in catchment areas was leading to the siltation of rivers and of the southern ports.[12] However, the overriding concern for all the Forest Officers was procurement of large amounts of timber. Even when Gibson was appointed as the Conservator of Bombay Presidency, his main task was to make provisions for the fulfillment of the State demand for timber, which is obvious from his reports of the Bombay Presidency.[13]

This claim regarding the introduction of scientific forestry has been defended by some writers like Marlene Buchy in his article, where he states that "...the scientific character of forestry in the 19 century is often incorrectly questioned in terms of the gaps in knowledge existing

---

11 See, Richard Grove, *Green Imperialism: Colonial Expansion, Tropical Island Edens and the Origins of Environmentalism 1600 to 1860*, CUP, First Indian Edition, OUP, 1995.

12 E. P. Stebbing, *Forests of India*, Vol. 1, London, 1922, pp. 120, 213, 215.

13 These reports give an ample description of the countryside of the Bombay Presidency, the density of the forests and their potentialities as a source of timber. See, Dr. A. Gibson, *Forest Report of the Bombay Presidency*, 1849–50 to 1855–56, Government Press, Bombay, 1957; Also see, Ajay Skaria, "Timber Conservancy, Desicationism and Scientific Forestry, The Dangs : 1840s to 1920s", *Nature and the Orient : The Environmental History of South and Southeast Asia*, (ed.), Richard Grove, V. Damodaran, Satpal Sangwan, OUP, Delhi, 1988, pp. 596–598.

at that time, which today have been filled by advances made in contemporary research. The fact that foresters were for a long time more concerned with trees rather than the forests or with the elements rather than the system, demonstrates a stage of knowledge which has in the mean time been surpassed to an extent, but it does not represent an absence of reflection".[14]

Keeping this argument in the background, it will not be out of place to make note of the orders issued by the medieval Maratha ruler, Shivaji to his naval staff, in which he clearly states that "a tree takes very long to grow. People look after them like their own children. Hence, do not cut them and cause misery to the people..."[15] These were the thoughts expressed by him when land was available in plenty. Extension of agriculture was the major concern of the rulers and deforestation was not an issue yet felt.

There is enough material to show the awareness and the lore for trees in pre-British Maharashtra. *Vrikshayurveda* (The Science of Plant Life by Surapala) was composed in sixth century AD. It enumerates several plants and the treatment they require for rapid and healthy growth.[16] They had knowledge of science of plants and they stated the importance of plants as "Ten wells are equal to one pond. Ten ponds are equal to one lake. Ten lakes are equal to one son. Ten sons are equal to one tree."[17] The sacred groves, known as *Devrais* in Konkan, are yet another example of conservation of forests in the pre-British period. These examples only show that traditional ethics had a positive bearing on the ecosystem in the pre-British system. It is not known

---

14 Marlene Buchy, "British Colonial Forest Policy in South India: An Unscientific or Unadopted Policy?", *Nature and the Orient: The Environmental History in South and Southeast Asia,* (ed.), R. Grove, V. Damodaran and S. Sangwan, OUP, Delhi, 1998, p. 637.

15 Ramachandra Pant Amatya, *Ajnapatre: Shivaji Rajanchi Rajneethi,* (Marathi), Poona, 1916, p. 38.

16 Surapala's *Vrikshayurveda,* (The Science of Plant Life), *Agri-History Bulletin,* No. 1, Asian Agro History Foundation, ICRISAT, Secunderabad, 1996, p. 37.

17 *Ibid,* p. 1.

whether the customs and traditions consciously imbibed ideological wisdom, and hence prevented people from exploiting natural resources mindlessly. But one can definitely see that these customs taught people to respect nature. They did not spread the culture of conquering nature.

In contrast to this pre-British tradition, the Forest Department of the British undertook forest operations in India on the basis of so-called scientific knowledge of forestry, though it was invariably the commercially valuable species which were looked after and preserved, consequently ignoring the subsistence needs of the poor lot. This was done by passing of different Forest Acts, providing the required legitimacy to the State. All along it was claimed by the British government that forests were preserved for people's welfare, for their own good. But in practice, it was mainly the imperial interests which were being looked after or rather given first priority by the Forest Department. The fulfillment of war demands (during the World Wars) is a glaring example of exploiting forests for imperial needs. Naturally, for the role it played, the Forest Department was praised by knowledgeable Forest Officers like E. P. Stebbing, who openly appreciated the department by saying, "If the Indian Civil Service was the steel frame of British Administration, the wooden frame of Forest Service proved to be equally tough and resilient during the War period."[18]

Curiously enough, the British officers who always complained that the Indian forests were badly destroyed by the Indians due to their destructive ways of exploitation, not only justified but praised the ruthless exploitation by the British State during the two World Wars. All this was done at the cost of interests of the people. In fact, the most far reaching and deeply adverse impact of British forest policies was felt by the communities, who had been depending on forests for their livelihood. Forest dwelling communities like the *Kolis* or the nomadic herders in the past offered a range of services to the rulers, and yet maintained their autonomy in the jungles. They lived on the fringes of society and maintained relations with village

---

18 E. P. Stebbing, *Forests of India*, Vol. 4, London, 1926, p. 147.

communities to a certain extent. Rulers remained content with revenue, rent and tributes paid by these communities. For instance, the *Kolis* of Bawan *Mavals* played a crucial role in Shivaji's struggle for *swaraj*. The political flexibility of indigenous rulers was part of an arrangement that was often, if not always, mutually beneficial. The limits to the power of the pre-British rulers, as far as these forest dwelling communities were concerned, were clear and the rulers did not try to change the lifestyle of these communities completely.[19]

However, the British attitude towards these people living on the fringes of society was different. The Agricultural Revolution in England and social changes in their society had led the British to tilt the scales heavily against the groups on the fringes of society in colonies like India.[20] A news item in the *Times of India*, Bombay, is a self-revealing example of this. The opening line of this item is "the *Pardhis* of Maharashtra, former hunters, are the hunted today". In 1872 the British made a law for controlling the activities of the tribes, which they identified as criminal. They set up 52 settlements or barbed wire ghettos throughout the country where these communities were forced to live. The *Pardhis*, traditionally a forest dwelling community, are one of the 42 so-called criminal communities in Maharashtra.[21] Due to their alleged criminal past the police and society in general, believed that *Pardhis* are born criminals.[22]

A survey conducted by Gadgil and Malhotra, "The Ecological basis of the Geographical distribution of the Dhangars: A Pastoral Caste Cluster of Maharashtra", concludes that the rapid destruction of forests witnessed during the last century in Maharashtra combined with over grazing has drastically reduced the carrying capacity of the land for

---

19 Mahesh Rangarajan, *Fencing the Forest: Conservation and Ecological Change in India's Central Provinces 1816–1914*, OUP, Delhi, pp. 196–201.

20 *Ibid*. The imposition of a more absolute notion of property complimented this process.

21 *Mahadev Kolis* was also one of the communities which was branded a criminal tribe out of the 42. The revolts of the *Kolis* have been discussed in chapter 7 of this book.

22 *Times of India*, Mumbai, 6 November 1997.

animal herds of the pastorals. The total amount of land available for grazing their animals has also shrunk with extension of cultivation by the rest of the population. Many pastoral groups, therefore, can no longer sustain themselves on their traditional occupation of animal husbandry. Consequently, the number of animal herds of the pastorals have gone down and goat, the animal most adopted to degraded vegetation, has become a more popular component of their animal herds. Also, owing to the degradation of tree cover and grazing areas the pastoralists have taken to cultivation increasingly.[23]

The pastoralists and the *Dhangars* of weaver castes mentioned above met with the same fate as so many other tribal communities, who were dependent on forests for their livelihood. Deprivation of means of livelihood combined with the harassment by overzealous forest officials, many a times resulted in resistance from the different sections of society. The forest officials' eagerness to meet the timber requirements and the revenue targets of the government made them overlook the subsistence needs of the tribals. The lack of understanding of indigenous customs and traditions, in short, their way of life, made matters even worse.[24] Many a times the resistance was offered in the form of individual offences. However, revolts of tribals were not an uncommon phenomena during this period. To the extent that the resistance was offered in the form of revolts such as the revolts of the *Kolis* or the *Varlis*. In the words of David Hardiman, they could be characterised as "a direct conflict between two forms of economic morality: that of the capitalist class and the other of the poor. The revolts

---

23 K. C. Malhotra and Madhav Gadgil, "The Ecological Basis of the Geographical Distribution of the Dhangars: A Pastoral Class Cluster of Maharashtra", *South Asian Anthropologist*, 1981, Vol. 2, No. 2, pp. 49–59.

24 Over zealousness of Forest Officers had become a matter of concern for the government also. From the beginning, this tendency of the Forest Officers was criticised by the revenue officials. However, later we find several instances where even the government, that is the higher authorities showed these concerns. See, *Standing Orders of the Forest Department*, corrected upto 1910, Bombay, 1918, p. 294; *Report of the Royal Commission on Agriculture in India*, Bombay, 1928, reprint 1983, p. 260, 269.

occurred as a result of this clash."[25] Incorporated in these revolts were the notions of community, moral economy and honour which involved among other things, the defence of the community and their maintenance of age old rights over land and forest resources.

At the organised level, particularly in Bombay Presidency, the tribals and peasants could get support from the forward communities like the *Chitpavan Brahmins*, other intellectuals and also some members of the aristocratic class. Forward caste people hailing from the landowning class were also able to understand the problems of the local people.

In the later period, this forest issue was also used by different political parties as an important political tool against the foreign authorities. For example, the Congress used it by way of *Jungle Satyagraha*.

However, the process of alienation of tribals from the forests did not come to a halt even after independence. The processes of urbanisation and commercialisation initiated and constantly encouraged by the British administrators, which continued into independent India, altered the lives of the entire rural populace of the country. The tribals were no exception to this. Most of them had either taken to marginal farming or had become unskilled labour in the nearby towns.

The environmental crisis has become a matter of major concern during the last few decades. Studying the history of human interaction with nature can be of use in formulating resource management policies. By this one can avoid the mistakes committed by predecessors and move away from Eurocentric models of development.

---

[25] The concept of 'moral economy' stated by E. P. Thompson explains this view. "...A view of social norms and obligations, of the proper economic functions of several parties within the community, which when taken together can be said to constitute the moral economy of the poor. An outrage to these moral assumptions quite as much as actual deprivation, was the usual occasion for direct action." E. P. Thompson, "The Moral Economy of the English Crowd in the Eighteenth Century", *Past and Present*, Vol. 50, February 1971, p. 79, cited in David Hardiman, *Feeding the Banias: Peasants and Usurers in Western India*, Delhi, OUP, 1996, p. 270.

This study of the history of forest policies in Maharashtra is an endeavour to build up the knowledge base on past forest conservation practices and policies, so that it may contribute to the contemporary debates on the sustainable management and exploitation of natural resources.

Appendix - I

Sacred Plants of Bombay Presidency

Botany Gazeteer of Bombay Presidency Volume XXV
Government Central Press, 1886.

Sacred Plants

Plants mentioned in the religious books and used in religious ceremonies of the Hindus.

The Hindus hold the belief that the Gods inhabit or frequent all sweet-scented or flowering plants; as such are never haunted by evil spirits. Therefore, a good member of plants are worshipped or used in some of their religious ceremonies. The most important are the following:

i. *Butea Frondosa* (Palas)
ii. *Prosopis Spicigera* (Shami)
iii. *Bauhinia Racemosa* (Apta)
iv. *Calotropis Gigantea* (Rui)
v. *Achyranthes Aspera* (Agarah)
vi. *F. Glomerata* (Umbar)
vii. *Ficus Bengalensis* (Vad)
viii. *F. Religiosa* (Pipal)
ix. *Cynodon Dactylon* (Dub)
x. *Eragrostis Cynosuroides* (Kush)
xi. *Eugenia Jambolana* (Jambul)

xii. *Mangifera Indica* (Amb)
xiii. *F. Cordifolia* (Pair)
xiv. *Ocymum Sanctum* (Tulus)
xv. *Phyllanthus Emblica* (Aula)
xvi. *Musa Paradisiaca* (Kela)
xvii. *Aegle Marmelos* (Bel)
xviii. *Saraca Indica* (Jassundi Ashok)
xix. *Rbita Pepo* (Kohala)
xx. *Mus Sativus* (Kakdi)
xxi. *Anus Odoratissimus* (Keuda)
xxii. *Azadirachta* (Nim)

Butea frondosa (Palas)

The leaves of this plant are trifoliate: the middle leaflet is supposed to represent Vishnu, the left Brahma and the right Shiv. Hence, its worship is enjoined in *Chaturmas Mahatma*. It is used in the following three great ceremonies:

1. The leaves are used as platters on the occasion of the investiture of the sacred thread, when a particular part of the ceremony, called chewul (that is when the barber removes the last tuft of hair from the head of the child to be invested), is being performed.
2. The dry twigs under the designation of *samidhas* are used for the feeding of *hom* or sacred fire, in the ceremony which goes under the name of *nava grahas* performed to secure the pacification of nine planets on the occasion of *vastu shanti*, that is entrance into a newly built house or one acquired from a non-Hindu.
3. The stem is used as a staff on the day of *sodmunj*, a part of the thread ceremony.

Prosopis spicigera (Shami, Shamri)

The worship of this tree is enjoined in vratraj (selections from *Bhavishya Puran*) to be performed on the 10 of *ashwin shudhapaksha* (Dasara festival). This day is also called *vijaya dashami*, as Arjun took down the weapons he had left on this tree many years before, and fought one of his memorable battles with them on this day. The tree is worshipped to obtain pardon for sins, success over enemies, and the realisation of the devotee's wishes. It is believed that Shami was transformed into a Goddess as she pleased Ram. The dry twigs are employed as *samidhas* for feeding the sacred fire. The leaves as *patri* are used in the worship of Ganapati, being thrown at the feet of that God's idol.

Bauhinia racemosa (Apta)

This tree called *ashmantak* in Sanskrit, is worshipped instead of Shami if the latter cannot be found. A complete narrative of this legend is also to be found in *Sahadrikand*. In the villages of the Konkan where Shami does not grow, the ceremony of the worship of Apta is performed somewhat as follows.

Early in the morning, people dressed in rich clothes assemble together and in procession go outside the town to the Apta tree where a priest recites some prayers prescribed in the ritual, the chief man of the town cuts its bough followed by other people. These boughs they take along with them to the nearest temple, usually of a Goddess, worship her and make her an offering of a few of the Apta leaves, together with flowers of french marigold *tajetes patula* and blades of rice and corn.

On returning home, they first make an offering of the remaining leaves to their household Gods, then to their priests accompanied by some copper or silver coins and to their parents and guardians or to the elderly people in the house; and lastly to their friends, saying "take this gold"; for the Hindus believe that the leaves over which the *mantras* are recited are worth gold or are transformed into gold.

Calotropis gigantea (Rui)

It is used in various ceremonies both religious and those of time hallowed custom. When a Hindu is to marry a third time, it is believed that the third wife will soon die; in order to avoid such a calamity the man is first married to this tree, which is then cut down. This ceremony is believed to ensure longevity of the fourth, but really the third wife whom he now marries.

Achyranthes aspera (Agarah)

A very common, ash coloured shrub which is about three feet. In *Chaturmas Mahatma* this plant is mentioned as the transformation of Budh. In *Smiritisar Granth* it is ordered to be used as a toothbrush. In *vratraj* it is directed that on fifth *bhadrapad shudh*, females should wash their mouth with 108 pieces of this plant using them as toothbrushes; but in case of need the number may be reduced as low as 7, also to bathe 108 times keeping each time a leaf of this plant on the head. It is also used as *samidhas*.

Ficus glomerata (Umbar)

It is mentioned in *Chaturmas Mahatma* that *Shukra* (teacher of the demons) was transformed into this tree. Its stem is used as nominal prop for *mandaps* on the day of *madmuhurth* a ceremony performed a day or two before marriage. It is mentioned in *gurucharitra* that Guru or the God *Dutta* or *Dattatreya* lives on this tree forever.

Ficus bengalensis (Vad)

The reason for worship of this tree is mentioned in *Vrat Koumudi*. It says that Savitri, the wife of Satyawan, worshipped this tree and thereby got back her deceased husband.

Ficus religiosa (Pipal)

Chaturmas mahatma tells us that this tree is a transformation of the God's Guru and is termed *ashwath*. It is stated in this *grantha* that this tree is frequented by all the Gods, and it is therefore ordered that it should be watered, worshipped and *pradakshinas* should be performed around it.

Cinodon dactylon (Durva or Dub)

Chaturmas Mahatma says that this grass is the transformation of *Rahu*. On the day of *Durva ashtami* on eighth *Bhadra-pad shudh* Hindus are ordered to go to this grass, pray to it, pluck it up from the ground, bring it home and worship it along with Shiva. And pray to Goddess Durga to increase their family as this grass is very fertile and gregarious. The marriage of this plant with Vad is also ordered.

Ocymum sanctum (Tulus)

This is the holy basil, common in all Hindu gardens, temples and courtyards. *Chaturmas Mahatma* tells us that it is a goddess and it should be watered and worshipped daily. The worship of this tree and its marriage with God Krishna is also mentioned in the *granth*.

Musaparadisiaca (Kela)

In *Vratraj* it is ordered that females should worship this tree on the fourth of *Kartik shudh*, whereby their husbands are said to survive and their (husbands) life is lengthened. The stems of the plants laden with their long rachises of fruits, are invariably placed by the people at the entrance of their houses, during their marriage or other festivals, as appropriate emblems of plenty and fertility.

Aegle marmelos (Bel)

Its leaf is the symbol of the Hindu triad and represents Brahma, Vishnu and Mahesh. It is employed as *patri* in worshipping the deity Shiva.

Saraca indica (Jasundi, Asok)

This is a small handsome tree. It is common in southern Konkan. *Vratraj* orders worship of this tree on *Chaitra shudh* 13.

Cucumis sativus (Kakdi)

In *Vratraj* it is mentioned that Rishis and Shiva told his wife Parvati to worship this tree, as by doing so females do not lose their husbands or at least they survive the wives. The fruit of this tree is cut into thin slices and employed in the worship of snakes in *shravan*.

Melia azadirachta (Neem)

When nectar was being taken to heaven from the world below, for the use of the Gods, it is believed that a few drops of it fell on this tree. Hence, on the New Year's day of *Shalivan Shaka* Hindus eat its leaves in the hope of acquiring freedom from disease.

Ipomaea pescaprae (Marayada vel or Marja vel)

The following curious practice is observed by Hindus at Bassein and in some villages of the Konkan, though there is no authority on this subject in India's sacred books; the creeper is twined around the cot of a Hindu mother on the sixth day after her confinement, and is intendent to serve as a protection to the newborn baby against any evil effects that might arise from the visit of Jewti or Satvi (Goddess of destiny.), who is supposed to come on this day for the purpose of writing the child's destiny on its forehead and sometimes even carry it away. The notion of this creeper being used in the manner above described, seems to be probably suggested by the fact that as it grows on the seaside and serves to bind the sands together, and prevents the encroachment of the sea, it might so to speak, be used to bind the child to its mother, and prevent the Satvi's evil intention.

Satvi is also propitiated by feasts (the members of the family keep a watch and pass the fifth and sixth nights in jollity and worship). This worship is of two kinds: one consists of drawing on the wall a female picture over whose face another face, made of dough is stuck and in whose four hands are four cradles with a baby in each. The figure is dressed in a cloth dyed with turmeric, and beneath it is placed a stoke, the one used for grinding masala, covered over with leaves of Rui (*Calotropis gigantea*) having the figures of children drawn on them with red lead. Both the figures, on the wall and on the leaves are then worshipped, different kinds of flowers and sweetmeats (usually gram and sliced coconut) are offered to the Goddess. These sweetmeats are afterwards consumed by each and every member of the family and distributed among the friends. A widow in the family generally conducts this ceremony. The other form of worship conducted by the father of the child consists in placing on a four footed (*chavrang*) a piece of gold or silver of the size of a 4 *anna* piece with the female figure stamped on it and a handful of rice. These are then worshipped and the gold or silver piece is subsequently, tied around the neck of the child to guard it against any danger to its life. This ceremony is known as *jivti pujan*.

Melia azedarach (Nim)

The nuts are frequently strung and worn round the necks, and during the prevalence of epidemics of small pox etc., are suspended over the doors and verandahs.

Elaecarpus ganitrus (Rudraksh)

This plant is to be found on the higher ghats. The five grooved and elegantly tubercled nuts are worn in the form of necklaces by Shiv *bhaktas* in order to obtain *sivlocke* that is, heaven wherein God Shiv resides and to gain his graces. A good account relating to this practice is to be found in *Sivlilamurt*. The plant being rare on the ghats, the nuts are imported in large quantities from Singapore, Nepal etc.

Ocymum sanctum (Tulus)

The *Vishnuvas* wear necklaces made of the root or stem of this shrub. Many plants besides the above are ordained by the sacred books of the Hindus to be used in religious ceremonies.

On Saturdays of the *Shravan* month, the Hindus prepare a kind of *bhaji* (called *panch bhelli bhaji*) of the following five plants.

i. *Cassia tora* (Takla)
ii. *Bauhinia malabarica* (Koral)
iii. *Amaranthus oleraceus* (Mat)
iv. *Celosia argentea* (Kurdu)
v. *Phalangium tuberosum* (Kuli)

Mat is always taken, even if one or more of the others are wanting. This kind of food is partaken of, because it is believed to be particularly acceptable to the Gods.

Eclipta prostata (Maaka)

It is used in worship on the days when *shrad* (office for the dead or ceremony in honour of the dead) is performed. The leaves of this plant and of *Ocymum sanctum* are thrown over the ball of cooked blessed rice (*pind*) made in honour of the dead. Some people also use the leaves and flowers of Augusta (*Sesbania grandiflora*).

A branch of Gometti (*Melothria heterophylla*) is suspended by some people over the hood of the idol of Cobra on *nag panchami* day, or during his worship.*

---

* These notes were drawn up from information supplied to the author by an intelligent Brahmin, and confirmed by his Hindu friend acquainted with the subject.

A Brief Description of the Daily Devotion

The Gods after being bathed (literally) in holy water or water over which certain *mantras* are recited and dried with a towel are placed in a copper dish, according to their piously supposed rank. By some they are kept on *chavrang*. Then *gandh* (sandlewood rubbed on stone with water) is applied to their foreheads followed by *akshata* (blessed rice) which is pressed or stuck over the *gandh*. Then flowers of various hews are thrown over the head of each image taking care to offer, whenever possible, those flowers which are most acceptable to each God. Then the images are dusted over successively with a pinch or two of *abhir, gulal, halad, kunku* and *shendur* (red lead). At the end of this *arti*, a lighted lamp fed with ghee is shown to or carried over the heads of the Gods and a food called *nayaved* is offered to them, the whole ceremony is being wound up by burning a small bit of camphor and *udbatti* or *agarbatti*. During the time the ceremony is being performed, both the hands and the mouth are constantly engaged.

In the Hindu worship described above, the images of the following five Gods are indispensibly present namely, Shiv, Vishnu, Ganapatti, Surya and one Devi named Annapurna, an incarnation of Parvati, the supplier of food. These five Gods form the *panchayatan*. To them are added the *penates*, or images of household Gods, varying in number, according to the wishes of the devotee.

Abhir is of brown colour and composed of various odoriferous substances.

Gulal is a reddish powder, made of the flowers of *Butea frondosa* called *kissu* in Gujarat. It is also made of sapan wood *Cesalpinea sapan* and starch obtained from the tuberous root of an aroid or some other plant.

Haladi is turmeric.

Kunku is the powder of the tuberous root of *Curcuma longa* coloured red by being immersed for some days in some alkaline solutions, generally of lime water. All these substances are of course, pulverised before being used. *Haladi* and *Kunku* are sprinkled specially over the head of Devi; and *Shendur* specially over Ganapati.

Nayaved may either consist of sugar alone or milk and sugar or any cooked food. When only of sugar or milk, it is offered in a brass cup, or of any other metal; but when it consists of any cooked food, it is offered in a brass dish only. This dish is placed on the ground in a square, previously marked out with the fingers dipped in water. The worshipper who during the whole ceremony squats on a very low stool, then takes two leaves of Tulus in his right hand, dips them in water, throws one on the food and the other, after five peculiar motions of the hand, on the Gods, keeping his eyes closed with the left hand.

Camphor is the concrete oil of *Cinammomum camphora*, a plant of China, Japan, Cochin China, etc. now introduced into Java.

Udbatti is composed of several odoriferous substances, the chief ingredient being benzoine, a balsamic exudation procured by making incisions into the bark of a tree named *Styrax benzoin*, indigenous in Borneo, Java, Sumatra and Siam.

Agarbatti is also composed of several substances, the chief of which is Aloes or eagle wood (*Aquillaria agallocha*), a tree found in Bengal.

All the ingredients composing these two *battis* are made into a paste, and then rolled into thin sticks, generally of the thickness of a small quill and then dried. Either one or the other is used, according to the means and fancy of the devotee.

In the selection of the flowers to be offered to the devotees colour is carefully attended to. During the worship of the deity Shiv white flowers are offered (those of water lillies *Nelumbium* and *Nymphoea* named Kamals being preferred); other colours are considered inadmissible. The leaves of bel tree are also offered; the whole leaf, and not the leaflets, must be used. No other God is worshipped with these leaves.*

Pandanus odoratissimus is cursed by Shiv, so its flowers though acceptable to the other Gods, are not used in his worship except on one particular occasion.

Vishnu receives either white or red flowers or those of any other colour. He is the only God who is worshipped with Tulus.

---

* In Chapter 18 of *Chaturmas Mahatma*, it is stated that other Gods may also be worshipped with these leaves, but it is never done.

Flowers of all colours are offered to Ganapati but red are preferred specially those of *Nerium odorum* (kanir) and *Hibiscus rosasinensis* (jasvand); yellow and white varieties are not rejected. During his worship Durva is absolutely necessary.

It is believed that white and yellow flowers are disliked by God Surya, and red are preferred. Hence, the red varieties of *Nerium odorum* and *Hibiscus rosasinensis* only are selected for offering. It may be for this reason that, in the gardens of Hindus, the red varieties of trees above mentioned are usually to be found. On some occasion Durva is offered to him. All flowers are offered to Devi. *Kunku* and *Haladi* are particularly placed over her head.

Goddess Gaur is much pleased with yellow flowers of *Thespesia Populnea* (bhendi). Hence, these are invariably offered to her. The same flowers are also offered to Ganapati and other deities male and female, who are not particularly averse to yellow.

Shank is worshipped with *gandh* and flowers but no rice is put over the *gandh*.

Leaves and flowers of *Callow tropis gigantea* are sacred to Hanuman and are used in his worship.

Euphorbia nerifolia (Thor) is sacred to the Goddess of Serpents, Manja.

Goddess Lakshmi resides or has her abode in the central yellow part (stamens and stigmas) of *kamal* (lotus). But any flower can be used in her worship.

The spiral vessels which abound in the leaves and scapes of the water lillies, after being carefully gathered, are made into wicks for lamps, which are lighted by the Hindus in the shrines of their idols.

The five flowers that tip the five arrows of Kamadev, the Indian Cupid, are: *Michelia* (champa), *Mesua ferrea* (nagkeshar), *Pavonia odorata* (Bala), *Mangifera indica* (ambi) and *Pandanus odoratissimus* (keoda).

Nyctanthes arbortristic (parijatak) is a tree supposed by the Hindus to have been brought from Heaven by God Krishna for his wife Satyabhama, on account of fragrance of its flowers.

Crotolaria Juncea Dalz.Gibs Bby.Fl.54.sunn, tag.

The great Hindu law giver Manu enjoined on the Shkatrias the use of the sacred thread of sun. It appears that Manu being a *Brahman*, always tried to keep this distiction, and claim superiority for his class. But, now a days, the sacred threads of almost all Hindus are made of cotton.

The common sacrificial woods of the Hindus on this side of India are:

i. *Butea frondosa* (Palas)
ii. *Prosopis spicigera* (Shami)
iii. *Calotropis gigantea* (Rui)
iv. *Achyranthes aspera* (Agarah)
v. *Ficus glomerata* (Umber)
vi. *Ficus bengalensis* (Vad)
vii. *Ficus religiosa* (Pipal)
viii. *Cynodon dactylon* (Dub)
ix. *Eragrostis cynosuroides* (Kush)

The five leaves (*panch pallav*), used by the Hindus in Konkan region of India as platters and for pouring libations, are gathered from the following trees:

x. *Mangifera indica* (Amb)
xi. *Eugenia jambolana* (Jambul)
xii. *Ficus bengalensis* (Vad)
xiii. *Ficus cordifolia* (Pair)
xiv. *Ficus religiosa* (Pipal)
xv. *Lea macrophyllal* (Dinda): The large leaves of this plant are used as platters for food every monday during the *shravan* month.

It appears that the Hindus are enjoined not to use tooth brushes made of coloured sap of plants; only those being lawful whose sap is colourless. *Smritisar Granth* gives a long list of plants twigs which can be used for the purpose. The following are a few of them which have been identified

i. *Bocagea dalzellii* (Sajiri)
ii. *Aegle marmelos* (Bel)
iii. *Feronia elephantum* (Kavit)
iv. *Zizyphus jujuba* (Bor)
v. *Mangifera indica* (Amb)
vi. *Butea frondosa* (Palas)
vii. *Acacia catechu* (Kair)

viii. *Albizzia lebbeck* (Shiris)
ix. *Anthocephalus cadamba* (Kadam)
x. *Calotropis gigantea* (Rui)
xi. *Wrightia tinctoria* (Kuda)
xii. *Achyranthes aspera* (Agarah)
xiii. *Briedelia montana* (Assana)

The fruits of Wagatti (*Capparis brevispina*) and of Gometti (*Melothria heterophylla*) Bryonia umbellata of Dalzell and Gibsis are eaten on *duadashis*, which occur in the month *Ashad*. The two fruits are invariably associated in the *bhaji* or dish made for those days.

Smithia sensitiva (kaola) is eaten in all seasons as a pot herb and is specially used on every Monday of *Shravan*.

Sacred Plant

Cassia tora (takla) is also eaten as a pot herb in Kolaba district (presently Raigarh) by some people.

Eclipta prostata (Maaka)

It is used in worship on the days when *shrad* (office for the dead or ceremony in honour of the dead) is performed. The leaves of this plant and of *Ocymum sanctum* are thrown over the ball of cooked blessed rice (*pind*) made in honour of the dead. Some people also use the leaves and flowers of Augusta (*Sesbania grandiflora*).

A branch of Gometti (*Melothria heterophylla*) is suspended by some people over the hood of the idol of Cobra on *nag panchami* day, or during his worship.*

* These notes were drawn up from information supplied to the author by an intelligent Brahmin, and confirmed by his Hindu friend acquainted with the subject.

Appendix - II

Minutes of Mr. Nulkar, Member of the Bombay Forest Commission 1887, recorded during the enquiry

In his minutes Nulkar said "When a strong and stable foreign government succeeds an indigenous one, in comparatively unsettled parts of the country, extreme measures of settlement at the earliest stages of the new administration are not frequently resorted to; but it is difficult to imagine a greater degree of oscillation of policy as regards the various agricultural interests, than is to be met within the different administrative measures taken at different epochs of the history of these districts, during the past 80 years. Beginning with the policy of putting every acre under cultivation and ending with the absorption of every inch that could be laid hands upon with impunity in the name of forest conservancy, a succession of extreme measures ran in opposite directions, inevitably resulting in the present chronic antagonism between the true interests of agriculture and a sound forest conservancy."

On the acquisition of these territories, their first want was found to be population, the country being for the most part a 'thinly inhabited forest.' The gradual settlement in it of the respectable and opulent natives of Mumbai and even the importation of Chinese emigrants were looked as possible means of bringing the land under cultivation. Even after half a century of British rule, waste land was so abundant as to create a feeling of despair as to the future of the district, and the increase of cultivation was so great an object of all administrative measures, that the poorest people were allowed to cut down, wherever they chose and as many trees as they liked for the purpose of cleaning the land. Around 1855–56 the revenue survey was introduced, its avowed policy being to assign as much land as could possibly be assessed as arable input under tillage. One of the recognised merits of the survey operations, was to show the largest possible area assessed as under cultivation. Waste lands such as *shindad, gurucharan* or forests which the people used freely for their agricultural and domestic wants were left almost untouched, because every other interest was to be made subservient to the great object of increased agriculture, and also because it was not thought worthwhile to go through the cost of measuring and mapping them along with the more important work to be done in connection with the lands under tillage.

Paripassu with these land revenue measures, spasmodic efforts were now and then made by the Forest Department, then in its embryo, to look after the well known Imperial forests and to collect such revenue as could be got under various names, based on custom – more or less modified – of former governments. But whenever these levies encroached upon the customary rights and privileges acquired by the rural communities, they were effectively opposed by the people who in 1851 are reported to have surrounded the Conservator in thousands, demanding immediate redress and "little disposed to stop short even of personal violence."

Along with this unsettled state of things as between agriculture and forest conservancy, there arose a fast increasing demand for timber and firewood from the rapidly growing city of Mumbai, in the immediate neighbourhood of which were suddenly added the almost insatiable requisitions of the two great systems of railway, which pierced the district in two directions. Thus, if the forests practically lay open to depredations by the railway contractors, it is no wonder that the poor inhabitants found themselves unable to resist the temptation of turning their opportunities of profiting by the windfall. But the wonder is that, while the short-sighted policy of the latter in denuding their own holdings of trees for small present profits has recoiled on them, the wholesale destruction of public forests along the lines of the railways by an authorised agency has scarcely been alluded to by the Forest Department, as a subject of complaint in the course of this enquiry.

Measurement and nominal record of the *shindad* lands, survey demarcations of some of the forests into Imperial and village shares being undertaken and partially carried out in some parts, subsequent to the revenue survey and settlement of 1855–56, resulted in a more or less restriction of the unlimited use, previously enjoyed by the people all over the wastelands. Between the first revenue survey and the year 1862 various orders, many of them contradictory, were passed with the view of gradually restricting the rights and privileges of the people regarding forest produce. The reorganisation of the forest establishments which took place in 1862 led to further, though still desultory, interference with and loud complaints from the people. The Forest Committee of 1863 and one or two subsequent ones, failed to bring about any satisfactory settlement of the varied questions which had cropped up. There followed louder complaints which pressed for settlement and increased with the strength of the grip of the newly augmented Forest Department which apparently felt the necessity for a financial justification for its existence, if not for further extension. The greed of forest revenue gathered such strength as to lead to the sale by auctions to strangers, by groups of villages, of every piece of grassland therein, sometimes including patches of waste around cultivated fields. While this ingenious raising of funds for the credit side of the forest budget often yielded not more than a rupee or two per hundred acres, the hardship to the villagers and their attitude towards such forest conservancy can be easily imagined.

At this juncture the Government of India passed its Forest Act VII of 1878 – an event which happened to be synchronous with the advent of presidency of a regime characteristic of its grasp, conception and vigour of execution of large measures of development and reform; a regime which would brook no human obstacle or opposition, which sought to compress within its narrow span of 5 years lease of office the ordinary achievements of a generation or two. It was under such a regime that Section 34 of the said Act (VII of 1878) was enforced in this presidency. The Section provides that "Within 12 months from the date on which this Act comes into force, within the territories administered by any local government and private persons in all forest lands or wastelands, then under its executive control for purposes of forest conservancy, determine which of such lands (if any) can, according to justice, equity and good conscience be classed as reserve forests or protected forests, under this Act, and declare by notification in the local official gazette any lands so classed to be received or protected forests as the case may be."

It seems plain that such lands may already be under the actual executive control of the local government for purposes of forest conservancy, and were alone to be notified and treated as protected or reserved forests. About 135 square miles had been upto the date of the Act of 1878, marked off and set apart as government forests in Thana and only this area could be correctly said to be under government control for purposes of forest conservancy, as contemplated by that Act. But by notification of 1 March 1879, the government arbitrarily added to this area about 1,529 square miles and suddenly, declared the same as government forest, in the name of law, justice, equity and good conscience. In short, by this notification, government claimed as forest every acre of land which did not happen to be entered in their land registers as actually owned and occupied by private individuals.

This sweeping measure, followed by the opinion of the Forest Committee of 1882 which was accepted by government, made *tahal rab* as not absolutely necessary for rice cultivation in Thana, and by decisions of Forest Settlement Officers declared the new forest areas to be gradually free of all private rights, reduced the people to straits unparalleled in the history of the fiscal administration of the Konkan. The principles which guided the new forest demarcation differed totally from those which regulated all former similar operations, in which a suitable portion of the forests was invariably set apart for village use as a first charge. Such a provision ceased to be a factor in the new demarcations. It was stated before the Commission that in some places fuel became so difficult to get even for cremation of the dead, that a man was compelled to pull down his own hut in order to use the timber for the purpose and some others had to bury their dead. The result of all the sharp practices resorted to by forest subordinates in the name of effective forest conservancy was, that the people were reduced to a choice between starvation and bribery to get what was their customary right. That they chose the latter as a temporary relief is now a matter of contemporary history. A huge system of bribery of the foresters by subscription in every village and hamlet, gradually came into existence. The people waited in apparent peacefulness and patience in hope of ultimate success of the constitutional means, they were advised to adopt in order to obtain justice. Most reliable evidence of this systematic bribery and corruption of forest servants would have poured in before the commission almost ad libitum, but for the caution which had to be administered to the witnesses that their statements were not privileged and might render them liable to criminal prosecution of defamation.

This temporary relief so universally purchased, could not last long. But for the tentative measures adopted by the government from time-to-time in order to relax the stringency of the new policy, the people would have been driven to the last resort. The antagonism and strife which, however, prevailed until very recently due to constant friction between the people and the conservancy department – a friction into which the executive of the country were irresistibly drawn – was the necessary result of the new born zeal for forestry, which knew no bounds. The administration of such homogenous parts of government as agriculture and forest conservancy had fallen into the hands of two distinct departments, one of them suddenly becoming the favourite of the prevailing policy of the hour. It is no wonder that the latter should become almost unruly. Complaints, accusations, recriminations between the people and the public departments embittered the ordinary official relations between the servants of the same government, and bewildered the people as to who really represented in the district the authority of the Central Government.

The people were prosecuted for trivial and technical offences while it was generally believed, not always without good grounds, that servants and contractors of the Forest Department escaped punishment even for grave offences. A forest guard who robbed cash from a poor man's hut on pretence of search for stolen forest produce and was sentenced to a term of imprisonment, found means to engage two professional lawyers to defend him. On leaving the prison, he easily found re-employment in the government forest service in another division of the same district, with the knowledge of of the high officers. A consequence of the effective combination to connive among the subordinate grades, brought about general demoralisation. The worst that could be alleged against the people was that they denuded their wants. But the subordinate foresters, freely blackmailed the people and the forest contractors carried on wholesale illicit cuttings of valuable timber in broad daylight without hindrance, the foresters sometimes actually helping them and the village authorities often impotently looking, having no correct knowledge of the nature of the proceedings.

The feeling of likes and dislikes, prejudice against the people as the greatest enemies of forest conservancy and confidence in their own subordinates, rose to such a degree as often to blind the highest officers to what was actually passing around him. Every forest offence was believed to be due to the connivance if not due to active assistance of the villagers. One of the forest departmental witnesses produced before the Commission in the original vernacular proceedings of enquiry, spoke of extensive robbery of teak timber by a forest contractor, as an example of how the villagers help or conceal such offences. But what did these very papers disclose? Facts quite the reverse of the allegations. The illicit cuttings were going on, wood cutters, carts and boats being employed openly on specific terms to carry the timber to Mumbai. The *Patil* failed to obtain explanation from the local forest guard as to the doings of the contractors. The first discovery and report of this systematic robbery was made by the *Talati* of the circle. He held a preliminary enquiry on the spot in the presence of the *Patil* and other village nobles. A complete chain of evidence was recorded against the offenders, several hundred pieces of freshly cut teak timber from the contractor's yard were attached. All the papers were sent to the *Mamlatdar*, earnestly requesting to be allowed to conduct the criminal prosecution against the offending contractors. Forwarding the papers on to the local European Assistant Forest Officer, the *Mamlatdar* stated that such offences had become too common, and therefore it was necessary that this case should be promptly dealt by him (the European Assistant Forest Officer). Three European officers happened to be encamping in the *taluka* at that time, the Conservator, the Divisional Conservator, the said Assistant Officer. Yet none of them considered it necessary, either to proceed to the scene of this wholesale plunder of public property or to hand over the case for trial to a competent magistrate. An inexperienced, new sub-assistant of the department was entrusted with a preliminary enquiry. As a result, the rich offenders got a good start in these new proceedings against them. Following this, some of the men withdrew their statements. Only a portion of those who had deposed before the *Talati* were questioned, the *Talati* not sent for, and after such a perfunctory enquiry, it was decided to give up the prosecution and deal with the case departmentally. In this manner, justice was allowed to miscarry so flagrantly in such an admittedly typical instance of systematic crime.

The entire tone of the forest administration of Thana, disclosed through the writings and proceedings of the head of the circle, was so stubbornly one-sided that there was small hope of an effective cure by conversion. Consistently with the popular rights and privileges, which have grown with the settlement and progress of the district, which are absolutely necessary for the continued well being of a rural population, and which the people have uninterruptedly enjoyed with the willing consent of the present and former governments.

Our recommendations, however, happen to be so opposed to what might be called the articles of faith embraced by the School of Forestry which supervises the conservancy in Thana, that I for one entertain no hope of securing anything like cordial cooperation on the part of any one who prides himself as the apostle of that faith, and who allows himself to fall into such extravagances as to prophecise in all seriousness of official reports, the utter destruction of the Bombay harbour as among the least evils, if his scheme for a wholesale adoption of the European continental systems of forest conservancy be not at once adopted here."

Bibliography

Primary Sources

Poona Archives – Peshwa Daftar, Poona

The Peshwa Daftar contains a total of 34,972 bundles, each containing 500 to 2,000 lose documents or a number of folded packets. These are available mostly in Marathi (Mood script) and some in Persian.

Dharmadhikari, Avinash (ed.), *Chitnisi Daftaratil Kagad Patranchi Varnanatmak Suchi,* Peshwa Daftar Abhilekhmala, Varnanatmak Suchi, Vol. 1–3, (Marathi), Government of Maharashtra, Mumbai, 1991.

Khobrekar, V. G. (ed.), *Parasnisi Daftaratil Kagad Patranchi Varnanatmak Suchi,* Kolhapur Abhilekhmala,Vol. 1 and 2 (Marathi), Government Press, Nagpur, 1975.

Kunte, B. G. (ed.), *Peshwe Sannika; Samshodhanasathi Niwadalale Kagad,* 2 Vol. (Marathi), Government Press, Nagpur, 1975.

Alienation Office Collection–English Records (East India Company).

Bharat Itihas Samshodhak Mandal, Poona

Banhatti, N. (ed.), *Ajnapatra,* (Marathi), Vividh Prakashan, Poona, 1974.

Kelkar, N. C., and, Apte, D. V., (ed.), *English Records on Shivaji 1659–1682,* Shivacharitra Karyalay, Poona, 1931.

Shivacharitra Sahitya Khand (Marathi), Vols. 1–13, Vividh Prakashan, Poona, 1926, reprinted in 1983.

Sri Ramdas Ani Ramdasi Granthamala: Sri Ramdasin Aitihasik Kagadpatre, Dhule, 1930.

Chimnaji Wad, Ganesh (ed.), *Selections from the Satara Raja's and Peshwa's Dairies*, Vols. 1–9, (Marathi), Bombay/Poona: Poona Deccan Vernacular Translation Society, 1906–11.

Maharashtra State Archives, Bombay

Agricultural Customs of the Thana district, Government of Bombay, 1896.

Annals of Indian Administration in 1870–71, Calcutta, 1871.

Annual Administrative Reports of the Forest Department of the Bombay Presidency, 1849–1945.

Annual Administrative Reports of the Rural Development in the Province of Bombay for the years, 1939–40 and 1940–41.

Annual Report on forest utilisation activities and economic research in Bombay State, 1949–1950, Government of Bombay, 1951.

Annual Return of Statistics Relating to Forest Administration in British India for the years 1914–15, 1916–17 and 1917–18.

Baillie, W. R., *Summary of Papers relating to the Madras and Bombay Forests*, Bengal Secretariat, Military Orphan Press, 1851.

Bombay – 1927–28: A Review of the Administration of the Presidency, Government Central Press, Bombay, 1929.

Bombay Forest Commission Report, Vol. 1–4, 1887.

Brandis, D., and Smythies, B. A. (ed.), *Report of the Proceedings of the Forest Conference*, Simla, 1875, Calcutta, 1876.

Brief Memorandum on the Material Condition of the People of the Bombay Presidency, 1881–1891, Government Press, Bombay.

Forest Department Code, The Indian Forest Code: General Regulations with Forms and Appendices, Superintendent Government Printing, Calcutta, 1913.

Forest Grievances Committee Report, Government Press, Bombay, 1926–27.

Forest – Anti-Forest Policy of the Bombay Government, Government of Bombay, 1899.

Gibson, Alexander, *Forest Report of Bombay Presidency*, 1849–50, 1850–51, 1851–52, 1852–53, 1853–54, 1854–55, 1855–56, 1856–57, 1857–58 and 1858–59, Maharashtra State Archives, Government Press, Bombay, 1961.

Howard, S. H., *Imperial Council of Agricultural Research Farmers and Forests*, reprinted from Indian Farming, Vol. II, No. 10, 1941.

Proceedings of the meeting of the Bombay Provincial Rural Development Board, Government Press, Bombay, 1942.

Report on the work of the Provisional Fodder and Grazing Committee, Bombay Presidency, 1940–41.

Report of the Forest Research in India, Provincial Reports, Bombay Presidency, 1925–26, 1926–27, 1928–29, 1935–36 and 1938–39.

Report of the Forestry Committee, Government of Bombay, 1929.

Revenue Department Files of the Bombay Presidency, Vol. 151, 1888 and Vol. 48, 1894.

Selections from the Records of the Government of India, (Home Department), Minute by the Governor General of India, Vol. 14, 28 February 1856, Calcutta Gazette Office, Calcutta, 1856.

Some Agricultural Customs of the Thana District, Government of Bombay, Bombay, 1896.

Standing Orders of the Bombay Presidency, corrected upto 1910, Government Press, Bombay, 1924.

Strettell, George, W., *A New Source of Revenue for India*, Marlborough and Company, London, 1878.

The Bombay Forest Rules, Government of Maharashtra, 1959.

The Fodder Problem in the Bombay Presidency, Government Press, Bombay, 1914.

The Work of the Forest Department in India prepared under the direction of the Inspector General of Forests to the Government of India, Imperial Forest Research Institute, Dehradun, April, 1920.

Quinquennial Review of Forest Administration in British India for the period 1909–10 to 1913–14, Government Press, Simla, 1914.

Quinquennial Review of Forest Administration in British India for the period 1914–15 to 1918–19, Government Press, Simla, 1920.

Working Plans of the Forest Department of the Bombay Presidency, 1902, 1905, 1908, 1910, 1914 and 1920.

Servants of India Society Library, Poona

Report of the Public Service Commission, Calcutta, 1886–87, printed by Superintendent of Government Printing, India, 1888.

Memorial from Poona Sarvajanik Sabha and the inhabitants of the city and camp of Poona, 3 March 1878, available at Gokhale Hall Library, Poona.

National Archives of India, New Delhi

Brandis, D., *Suggestions Regarding Forest Administration in Madras Presidency*, Madras, 1883.

Legal Department, A, March 1878, No. 43 to 142.

"Minute By the Most Noble The Governor General of India", *Selections from the Records of the Government of India, Home Department*, Calcutta Gazette Office, Calcutta, 1856.

Proceedings of the Forestry Conference (held on the 9 and 10 December, 1937), New Delhi, 1937.

Revenue and Agriculture Department, Forest Branch Proceedings for March 1878, No. 54, Part B.

Report of the Royal Commission on Agriculture, first print, Bombay, 1928, reprint, New Delhi, 1987.

Report from the sub-committee of the Poona Sarvajanik Sabha, appointed to collect information to be laid before the East India Finance Committee on matters relating to India, Poona, 1873.

"Report on the Teak Forests of the Tenasserim Provinces", *Selections from The Records of The Bengal Government*, Military Orphan Press, 1852.

Rules for the leases or sale of waste lands in India, compiled in the Revenue and Agriculture Departments of the Government of India, 1892–1904.

Statistical Report of the Kolaba Agency, Government Press, Bombay, 1851.

Working Plan Report of Fuel and Fodder Reserves of the Ahmednagar Division, Central Circle, Bombay Presidency, 1940.

Gazetteers, Census Reports, Magazines and Newspapers

Gazetteers

Imperial Gazetteer of India, Bombay Presidency, Nasik, Canara, and Kolaba–Janjira Districts, Government Press, Bombay, 1883.

Imperial Gazetteer of India, Bombay Presidency, Khandesh, Government Press, Bombay, 1880.

Imperial Gazetteer of India, Bombay Presidency, Poona District, Government Press, Poona, 1885.

Imperial Gazetteer of India, Bombay Presidency, Thana District, Government Press, Bombay, 1882 and 1908.

Imperial Gazetteer of India, Bombay Presidency, Ratnagiri and Satara Districts, Government Press, Bombay, 1887.

Imperial Gazetteer of India, Botany Gazetteer by Campbell, Vol. XXV, Government Press, 1886.

Gazetteer of India, Maharashtra State, Botany and Flora of Maharashtra, Government Press, Nagpur, 1987.

Census Reports

Imperial Census Reports of India, Maharashtra Presidency, 1871, 1881,1891 and 1901.

Magazines

Journal of Bombay Natural History Society, Natural History Society, Bombay (1908–1913).

Quarterly Journal of the Poona Sarvajanik Sabha (English and Marathi).

Indian Forester.

News papers

Kesari (Marathi)

Times of India

Arunoday (Marathi)

Bombay Chronicle

Secondary Sources

Adas, M., "From avoidance to confrontation: Peasant protest in precolonial and colonial southeast Asia", *Comparative Studies in Society and History*, Vol. 23, 1981.

Agarwal, A., "Human-Nature interactions in a third world country", *The Environmentalist*, Vol. 6, 1986.

Agarwal, D. P., and Sood, R. K., "Ecological factors and the Harappan civilisation", *Harappan Civilisation*, (ed.), Possehl, G. L., Oxford and IBH, New Delhi, 1982.

Aiyappan, A., *Report on the Socio-Economic Conditions of the Aboriginal Tribes of the Province of Madras*, Madras Government Press, Madras, 1948.

Albion, Robert Greenhalgh, *Forests and Sea Power: The Timber Problem of The Royal Navy*, Harvard University Press, Cambridge, Massachusetts, 1926.

Alton, B., and Allchin, F. R., *The Birth of Indian Civilisation*, Penguin, Harmondsworth, 1968.

Ambedkar, B. R., *The Untouchables*, Amit Book Company, New Delhi, 1948.

Anonymous, *A Manual of Forest Law compiled for the use of the students at the Imperial Forest College, Dehradun*, Government Press, Calcutta, 1906.

Anonymous, *Report of the Forest Committee*, Vol. I, Government Press, Madras, 1913.

Anonymous, *Report of the Imperial Economic Committee: Tenth Report: Timber*, HMSO, London, 1920.

Anonymous, *Report of the Royal Commission of Agriculture*, HMSO, London, 1927.

Apte, B. K., *A History of the Maratha Navy and Merchant Ships*, State Board for Literature and Culture, Bombay, 1972.

Arnold, D., "Rebellious Hillmen : The Gudem Rampa Rebellions (1829–1914)" *Subaltern Studies I*, (ed.), Ranjit Guha and Hardiman, David, OUP, New Delhi, 1982.

Ashby, E., *Reconciling Man with the Environment*, Oxford University Press, Oxford, 1978.

Bachketi, N. D., *Social Forestry in India: Problems and Prospects*, Birla Institute of Scientific Research, New Delhi, 1984.

Baden-Powell, B. H., *Forest Law: Course of Lectures on the Principles of Civil and Criminal Law and on the Law of the Forests*, Bradbury Agnew and Company, London, 1893.

————, "The Political Value of Forest Conservancy", *Indian Forester*, Vol. 2, January 1877.

————, "Note on Demarcation of the Forest Area in Districts Containing Hill or Mountain Ranges", *Indian Forester*, January, 1877.

————, *The Forest System of British Burma*, Government Press, Calcutta, 1874.

————, *A Manual of Jurisprudence for Forest Officers*, Government Press, Calcutta, 1882.

————, *Memorandum on Forest Settlement in India*, Government Press, Calcutta, 1892.

————, "On the defects of the existing Forest Law (Law (Act Vii) of 1865) and Proposals for a New Forest Act", *Reports of the Proceedings of the Forest Conference, 1873–74*, (ed.), Baden-Powell B. H., and Gamble, J. S., Government Press, Calcutta, 1875.

Badgeley, C., Kelley, J., Pilbean, D., and Ward, S., "The Palaeobiology of South Asian Miocene Hominoids", *The People of South Asia*, (ed.), J. R. Lucas, Plenum Press, New York, 1984.

Baillie, W. R., *Summary of Papers relating to the Madras and Bombay Forests*, Bengal Secretariat, Military Orphan Press, Bengal, 1851.

Baker, C. J., *An Indian Rural Economy 1880–1995*, OUP, Delhi, 1984.

Baker, D. E. U., "A serious time: Forest *Satyagraha* in Madhya Pradesh, 1930", *Indian Economics and Social History Review*, Vol. 21, 1984.

Beckerman, W., "Economists, Scientists and Environmental Catastrophe", *Oxford Economic Fapers*, Oxford University Press, Oxford, Vol. 24, 1972.

Bennet, S., "Shikar and the Raj", *South Asia N. S.*, Vol. 7, 1984.

Benton, Ted, "Marxism and Natural Limits: An Ecological Critique and Reconstruction", *New Left Review*, November–December, 1989.

Best, J. W., *Forest Life in India*, John Murray, London, 1935.

Bhave, Vasudev, Krishna, *Peshwakaleen Maharashtra*, (Marathi), Indian Council of Historical Research, New Delhi, 1976.

Billings, W. D., *Plants, Man and the Ecosystem*, Belmont, 1970.

Birrel, J., "Common Rights in the Medieval Forest", *Past and Present*, No. 117, 1987.

Blair, H., "Social Forestry in India: Time to Modify Goals", *Economic and Political Weekly*, 21 July, 1986.

Bokeel, Milind, "Pranyanna Abhav, Mahsaana Vanvaas", (Marathi), *Saptahik Sakaleen*, Diwali edition, 1996.

Bloch, M., *French Rural History: An Essay on its Essential Characteristics*, Reprint 1931, Routledge and Kegan Paul, London, 1978.

Brahme, S., and Upadhya, A., "A Critical Analysis of the Social Formation of and Peasant Resistance in Maharashtra", Mimeo, Three Volumes, Gokhale Institute of Politics and Economics, Poona, 1979.

Brandis, Dietrich, *Forestry in India: Origins and Early Development*, Brigton, 1872, reprint by WWF India, Nataraj Publications, Dehradun, 1994.

—————, *The Distribution of Forests in India*, McFarlane and Erskine, Edinburgh, 1873.

—————, *Memorandum on the Forest Legislation Proposed for British India (other than the Presidencies of Madras and Bombay)*, Government Press, Simla, 1875.

—————, *Suggestions Regarding Forest Administration in the North Western Provinces and Oudh*, Government Press, Calcutta, 1882.

—————, *Progress of Forestry in India*, McFarlane and Erskine, Edinburgh, 1884.

—————, *Indian Forestry*, Oriental Institute, Woking, 1897.

Branford, H. R., "Regeneration with the Assistance of Tangy in Burma", *Indian Forest Records*, Vol. 11, Part 3, 1925.

Brara, R., "Shifting Sands: A Study of Rights in Common Pastures", *Mimeo*, Institute of Development Studies, Jaipur, 1987.

Brenner, R., "Agrarian Class Structure and Economic Development in Pre-Industrial Europe", *Past and Present*, 1976.

—————, "Dobb on the Transition from Federalism to Capitalism", *Cambridge Journal of Economics*, Vol. 2, No. 2, 1978.

Brennon, A., *Thinking About Nature*, Routledge, London, 1988.

Burman, Roy, K. J., "Sacred Groves", *Journal of Human Ecology*, Vol. 6, No. 4, 1995.

Calman, L., *Protest in Domestic India*, Westview Press, Boulder, 1985.

The State of India's Environment 1984–85: A Second Citizens' Report, Centre for Science and Environment, New Delhi, 1985.

Champion, H. G., and Osmaston, F. C., (ed.) E. P. Stebbing, *The Forests of India*, Oxford University Press, Oxford, Vol. 4, 1962.

Champion, H. G., and Seth, S. K., *General Silviculture for India*, Manager of Publications, Delhi, 1968.

Chandra, G. S., and Srivastava, S. S., "Forestry Economics in India, Some Areas of Research", *Arthavijnana*, Vol. 10, 1968.

Chandrasekhar, D.M., Krishnamurthy, B. V., and Ramaswamy, S. R., "Social Forestry in Karnataka: An Impact Analysis", *Economic and Political Weekly*, 29 September, 1987.

Chandramouli, K., "Kondapalli Through the Ages", *Archaeological Series*, Department of Archaeology and Museum, Government of Andhra Pradesh, 1968.

Chapekar, L. N., *Thakurs of Sahayadri*, OUP, Mumbai, 1959.

Chatterjee, P., "More on Modes of Power and the Peasantry", *Subaltern Studies II*, (ed.), Ranjit Guha, Oxford University Press, Delhi, 1983.

————, "Gandhi and the Critique of Civil Society", *Subaltern Studies III*, (ed.), Ranjit Guha, Oxford University Press, Delhi, 1984.

Chaudari, K. A., *Ancient Agriculture and Forestry in Northern India*, Mumbai, 1977.

Choksey, R. D., *Economic Life in the Bombay Konkan*, Pheonix Publication, Bombay, 1937.

Cleghorn, H., *Forests and Gardens of South India*, W. H. Allen, London, 1860.

Cohen, Y., *Man in Adaptation, The Cultural Present*, Aldine Chicago, Second edition, 1974.

Conklin, H., "An Ethno-Ecological Approach to Shifting Agriculture", *Environment and Cultural Behaviour*", (ed.), Vayda, A. P., Academic Press, New York, 1969.

Cronon, W., *Changes in the Land: Indians, Colonists and the Ecology of New England*, Hill and Wang, New York, 1983.

Crosby, A., *Ecological Imperialism: Biological Expansion of Europe*, Cambridge University Press, Cambridge, 1986.

D'Arcy, W. E., *Preparation of Forest Working Plans in India*, Government Press, Calcutta, 1910.

Dasgupta, P., *The Control of Resources*, Oxford University Press, Delhi, 1982.

Dasgupta, S., "Local Politics in Bengal: Midnapur District 1907–1923", Unpublished thesis, School of Oriental and African Studies, University of London, 1980.

Dasmann, R., "Towards a Biosphere Consciousness", Worster, D., *The Ends of the Earth: Perspectives on Modern Environmental History*, Cambridge University Press, Cambridge, 1988.

Desai, Sudha, V., *Social Life in Maharashtra under the Peshwas*, Popular Prakasan, Mumbai, 1980.

Dhareshwar, S. S., "The Denuded Condition of the Minor Forest in Canara Coastal Tract: Its History and a Scheme for its Regeneration", *Indian Forester*, Vol. 67, 1941.

Dharmadhikari, Avinash, (ed.), *Chitnis Daftar Kagad Patranchi Varnanatmak Suchi*, English Records on Shivaji, 1659–1682, Shivcharitra Karyalay, Poona, 1931.

Dhavalikar, M. K., *The First Farmers of the Deccan*, Deccan College, Poona, 1988.

Dienner, P., and Robkin, E., "Ecology and the Search for Cultural Origins: The Question of Islamic Prohibition", *Current Anthropology*, Vol. 19, 1978.

Digby, S., *War Horse and Elephant in the Delhi Sultanate*, Clarendon Press, 1971.

Dixit, K. R., assistant by Charulata Patil, *Maharashtra in Maps*, Maharashtra State Board for Literature and Culture, Bombay, 1986.

Dunn, F. L., "Rain Forest Collectors and Traders: A Study of Resource Utilisation in Modern and Ancient Malaya", Malasia branch of Royal Asiatic Society, Monograph No. 5, 1976.

Eaton, P., "Customary Land Tenure and Conservation in Papua New Guinea", *Culture and Conservation: The Human Dimension in Environmental Planning*, (ed.), NcNeely, J. A., and Pitt, D., Croom Helm, Dublin, 1985.

Eckholm, Erik, P., *Down to Earth: Environment and Human Needs*, Pluto Press, London, 1982.

Elliot, J. G., *Field Sports in India, 1800–1947*, Gentry Books, London, 1973.

Elphinstone, Mount Stuart, *Territories Conquered from Peshwa: A Report*, Calcutta, 1821, reprint New Delhi, 1973.

Elwin, Varrier, *The Baiga*, John Murray, London, 1939.

————, *The Agaria*, Oxford University Press, Calcutta, 1942.

————, "The Aboriginals", *Oxford Pamphlets and Indian Affairs*, Blackwell Publishing on behalf of Royal Institute of International Affairs, Oxford, No. 14, 1943.

————, "Saora Fituris", *Man in India*, Vol. 25, 1945.

————, *Leaves from the Jungle*, Reprint 1936, Oxford University Press, London, 1958.

————, *A Philosophy for NEFA*, Government Press, Shillong, 1960.

Engels; Frederick, *Dialectics of Nature*, Progress Publishers, Moscow, 1954.

————, *The Peasant War in Germany*, Progress Publishers, Moscow, 1956.

Enthoven, R. E., *The Tribes and Castes of Bombay*, Central Press, Government of Bombay, Vol. 1–3, 1920–1922.

Faith, R., "The Great Rumours of 1377 and Peasant Ideology", *The English Rising of 1381*, (eds.), Hilton, R. H., and Ashton, T. H., Cambridge University Press, Cambridge, 1984.

Fernandes, W., and Menon, G., *Tribal Women and Forest Economy: Deforestation, Exploitation and Status Change*, Indian Social Institute, New Delhi, 1987.

Fernandes, W., Menon, G., and Viegas, P., *Forests, Environment and Tribal Economy: Deforestation, Impoverishment and Marginalisation of Orissa*, Indian Social Institute, New Delhi, 1988.

Fernow, B. E., *A History of Forestry*, Toronto University Press, Toronto, 1907.

Food and Agricultural Organisation, United Nations, (FAO 1984), "Intensive Multiple-Use Forest Management in Kerala", *FAO Forestry Paper*, 53, Rome, 1984.

Ford Robertson, F. C., *Our Forests*, Government Press, Allahabad, 1936.

Fortmann, L. P., and Fairfax, S. K., "American Forestry Professionalism in the Third World – Some Preliminary Observations", *Economic and Political Weekly*, Bombay, 12 August 1989.

Freeman, M. R., "Grafs and Gaffs: A Cautionary Tale in the Common Property Resources Debate", *Common Property Resources: Ecology and Community Based Sustainable Development*, (ed.), Berkes, F., Belhaven Press, London, 1989.

Fukozawa, H., *The Medieval Deccan – Peasants, Social Status and States, 16 to 18 Century*, Delhi, 1991.

Gadgil, D. R., "Immigrant Traders in Poona in the Eighteenth Century", *Arthvigyan*, March, 1959.

Gadgil, Madhav, "Social Restraints on Resource Utilisation: The Indian Experience", *Culture and Conservation: The Human Dimension in Environmental Planning*, (ed.), McNeely, J. A., and Pitt, D., Croom Helm, Dublin, 1985.

————, "Cultural Evolution of Ecological Prudence", *Landscape Planning*, Vol. 12, 1985.

————, "Diversity: Cultural and Biological", *Trends in Ecology and Evolution*, Vol. 2, No. 2, 1987.

————, "The Indian Heritage of a Conservation Ethic", *Conservation of the Indian Heritage,* (ed.), Allchin, E. R., and Thapar, D. K., Cosmo Publications, New Delhi, 1989.

Gadgil, Madhav, and Guha, Ramachandra, *This Fissured Land: An Ecological History of India,* Oxford University Press, New Delhi, 1992.

Gadgil, Madhav, and Iyer, P., "On the Diversification of Common Property Resource Use by Indian Society", *Common Property Resources: Ecology and Community Based Sustainable Development,* (ed.), Berkes, F., Belhaven Press, London, 1989.

Gadgil, Madhav, and Malhotra, Kailash, "The Ecological Basis of the Geographical Distribution of the Dhangars: A Pastoral Class Cluster of Maharashtra", *South Asian Anthropologist,* Vol. 2, No. 2, 1981.

Gadgil, Madhav, Prasad, S. N., and Ali, Rauf, *Forest Management in India: A Critical Review,* Centre for Theoritical Studies, Indian Institute of Sciences, Bangalore, 1982.

————, "Forest Management and Forest Policy in India: A Critical Review", *Social Action,* Vol. 33, 1983.

Gadgil, Madhav, and Chandran Subhash, "Environmental Impact of Forest Based Industries on the Evergreen Forests of Uttara Kannada district: A Case Study", *Final Report Submitted to the Department of Ecology and Environment,* Government of Karnataka, 1989.

Gadgil, Madhav, and Vartak, V. D., "Sacred Groves of Western Ghats of India", *Economic Botany,* Vol. 30, 1976.

Gare, Govind, *Sahayadriteel Adivasi: Mahadev Koli,* (Marathi), Sahadhyayan Publishers, Bombay, 1974.

Gause, G. F., *The Struggle for Existence,* Hufuner Publishing Company, New York, 1969.

Geoffrey and Jellieoe, Susan, *The Landscape of Man, Shaping the Environment from Pre-history to the Present Day,* Thames and Hudson, London, 1975.

Gerald, Folay and Bashard, Geoffey, "Farm and Community Forestry", *Technical Report No. 3, Energy Information Programme Earthscan,*

International Institute for Environment and Development, Congress of USA, Office of Technology Assessment, 1984.

Gibson, Alexander, *Forest Reports of the Bombay Presidency 1849–50, 1850–51, 1856–57, 1857–58, 1858–59, 1859–60*, Education Society's Press, Bombay, 1861.

Gordon, Stewart, "Revenue System of the Marathas under Nana Phadnis", *Medieval Deccan History*, (ed.). Kulkarni, A.R., Nayeem, M.A., and De Souza, T.R., Popular Prakashan, Bombay, 1996.

Gottfried, Robert, S., *The Black Death, Nature and Human Disaster in Medieval Europe*, Hale, London, 1983.

Goudie, A. S., *The Human Impact on the Environment*, Basil Backwell, Oxford, Second edition, 1987.

Goulding, M., *The Fishes and the Forest, Explorations in Amazonian Natural History*, University of California Press, Berkeley, 1980

Government of India, *Report of the Forest Committee*, Government Press, Simla, 1929.

————, *India's Forests and the War*, Ministry of Agriculture, New Delhi, 1948.

————, *The National Forest Policy*, Government Press, Delhi, 1952,

————, *The Second Five Year Plan*, Planning Commission, Delhi, 1956.

————, *Report of the Scheduled Areas and Scheduled Tribes Commission*, Vol. 1, 1960–61, Manager of Publications, Delhi, 1961.

————, *History of Indian Railways*, Government Press, Delhi, 1964.

————, *Report of the Committee on Tribal Economy in Forest Areas*, Department of Social Welfare, Delhi, 1967.

————, *Task Force Report on Taking Forestry to the People*, Ministry of Agriculture, Delhi, 1980.

Grigg, D. B., *The Agricultural Systems of the World: An Evolutionary Approach*, Cambridge University Press, Cambridge, 1980.

Grigson, W. V., *The Maria Gonds of Bastar,* Oxford University Press, London, 1938.

Grove, Richard, H., "Colonial Conservation, Ecological Hegemony and Popular Resistance: Towards a Global Synthesis", *Imperialism and the Natural World,* (ed.), J. M. Mackenzie, Manchester University Press, Manchester, 1990.

Guha, Ramachandra, *Comparative Studies in Society and History,* Vol. 36, No. 3, Cambridge 1993.

————, "Forestry in British and post-British India: A Historical Analysis", *Economic and Political Weekly,* 29 October and 5–12, November, 1983.

————, *The Unquiet Woods: Ecological Change and Peasant Resistance in the Himalaya,* Oxford University Press, Berkeley, 1989.

————, "Radical American Environmentalism and Wilderness Preservation: A Third World Critique", *Environmental Ethics,* Spring, 1989.

————, *Green Imperialism: Colonial Expansion, Tropical Island Edens and the Origins of Environmentalism, 1600–1860,* CUP, First Indian edition, OUP, 1995.

————, "The Making of the 1878 Act: An Early Environment Debate", *The Indian Economic and Social History Review,* Vol. 27, No. 1, 1990.

————, "Writing Environmental History in India", *Studies in History,* New Series, Vol. 9, No. 1, 1993.

————, *Social Ecology,* Oxford University Press, Delhi, 1994.

————, "Dietrich Brandis and Indian Forestry : A Vision Revisited and Reaffirmed", *Village Voices, Forest Choices: Joint Forest Management in India,* (ed.), Poffenberger, Mark, and McGean, Betsy, Oxford University Press, Delhi, 1996.

Gupta, R., Bannerji, P., and Guleria, A., *Tribal Unrest and Forestry Management in Bihar,* Indian Institute of Management, Ahmedabad, 1981.

Gupta, T., Sambrani, S., "Control of Shifting Cultivation: The Need for an Integrative Approach and Systematic Appraisal", *Indian Journal of Agricultural Economics*, Vol. 31, 1978.

Halappa, G. S., *History of Freedom Movement in Karnataka*, Government Press, Bangalore, Vol. 2, 1964.

Hanumantha Rao, C. H., *Agricultural Growth, Rural Poverty and Environmental Degradation in India*, Institute of Economic Growth, Oxford University Press, New Delhi, 1994.

Hardiman, David., "Small-Dam Systems of the Sahayadris", *Nature, Culture, Imperialism: Essays on the Environmental History of South Asia*, (ed.), David Arnold and Ramachandra Guha, Oxford University Press, Delhi, 1995.

————, "Farming in the Forest: The Dangs 1830–1992", *Village Voices, Forest Choices: Joint Forest Management in India*, (ed.), Mark Poffenberger, Mark and McGean, Betsy, Oxford University Press, 1996.

————, *Feeding the Baniya: Peasants and Userers in Western India*, Oxford University Press, Delhi, 1996.

————, "Community, Patriarchy, Honour: Raghu Bhangre's Revolt", *Journal of Peasant Studies*, October, 1995.

Hardin, G., "The Tragedy of the Commons", *Science*, Vol. 162, 1968.

Harris, M., *Culture, People and Nature: An Introduction to General Anthropology*, Harper and Row, New York, 1980.

Hays, S. P., *Conservation and Gospel of Efficiency: The Progressive Conservation Movement 1880–1920*, Harvard University Press, Cambridge, Massachussetts, 1958.

————, *Beauty, Health and Permanence: Environmental Politics in the United States, 1955–85*, Cambridge University Press, Cambridge.

Hearle, N., *Working Plan of the Tehri Garhwal Leased Forests, Jaunsar Forest Division*, Government Press, Allahabad, 1888.

————, *Working Plan of the Deoban Range, Jaunsar Forest Division, Northwestern Provinces*, Government Press, Allahabad, 1989.

Heske, F., *German Forestry*, Yale University Press, New Haven, 1937.

Hivale, S. and Elwin, V., *Songs of the Forest: The Folk Poetry of the Gonds*, Allen and Unwin, London, 1935.

Hutchinson, J., Clark, G., Jope, E. M. and Riley, R., *Early History of Agriculture*, Oxford University Press, Oxford, 1977.

International Institute for Environment and Development and World Resources Institute, *World Resources, 1987*, Basic Books, New York, 1987.

Jarrige, J. F. and Meadow, R. H., "The Antecedents of Civilisation in the Indus Valley", *Scientific American* Vol. 243, 1980.

Jodha, N. S. "Common Property Resources and Rural Poor in Dry Regions of India", *Economic and Political Weekly*, 5 July 1986.

Jones, E. L., "The Environment and the Economy", Peter Burke, *The New Cambridge Modern History*, (ed.), Cambridge University Press, Cambridge, Vol. 13, 1979.

Kangle, R. P., *Arthashastra*, University of Bombay, Bombay, 1969.

Karve, Iravati, *Hindu Society: An Interpretation*, Deccan College, Poona, 1961.

————, *Yuganta. The End of an Epoch*, Sangam Books, Poona, and Orient Longman, New Delhi, 1974.

————, "The Tribals of Maharashtra", *Collected Papers of Iravati Karve*, reprinted by Urgent Research in Social Anthropology, Indian Institute of Advanced Studies, Simla, Vol. 10, 1969.

Kosambi, D. D, *Myth and Reality*, Bombay, 1962, reprinted in 1994.

————, *The Culture and Civilisation of Ancient India in Historical Outline*, Vikas Publishing House, Delhi, 1970.

Kosambi, Meera, "The Konkan Ports and European Dominance : An Urban Perspective", *Medieval Deccan History*, (ed.), Kulkarni, A. R., Nayeem, M. A., and de Souza, T. R., Popular Prakasan, Bombay, 1996.

Kothari, A., Pande, P., Singh, S., and Variava, D., *Management of National Parks and Sanctuaries in India: A Status Report*, Indian Institute of Public Administration, New Delhi, 1989.

Krishnakutty, C. N., and Chundamannil, M., "Are Eucalyptus Plantations Fulfilling the Goals in Kerala?", *Eucalyptus in India: Past, Present and Future,* Kerala Forest Research Institute, 1986.

Kumar, D., *The Cambridge Economic History of India,* Cambridge University Press, Cambridge, Vol. 2, 1983.

Kumar, Ravindra, *Western India in the Nineteenth Century: A Study in the Social History of Maharashtra,* London, 1968.

————, "Deccan Riots of 1875", *Journal of Asian Studies,* Vol. 24, August 1965.

Kulkarni, A. R., *Maharashtra in the Age of Shivaji,* Poona, 1969.

Lal, M., *Settlement History and the Rise of Civilisation in the Ganga-Yamuna Doab from 1500 BC–AD 300,* B. R. Publishing, Delhi, 1984.

Leathart, S., "Review of N. D. G. James, A History of English Forestry", *Times Literary Supplement,* 8 January, 1982.

Linebaugh, P., "Karl Marx, The Theft of Wood, and Working Class Composition: A Contribution to the Current Debate", T. Platt and P. Kakagi (eds.), *Crime and Social Justice,* Macmillan, New York, 1901.

Madleod, W. C., "Conservation among Primitive Hunting Peoples", *The Scientific Monthly,* December, 1936.

Mahalanobis, P. C., *The Approach of Operational Research to Planning,* Indian Statistical Institute, Calcutta, 1963.

Malhotra, K. C., "Socio-Biological Investigations among the Nandiwallas of Maharashtra", *Bulletin of the Urgent Anthropological and Ethnological Sciences,* No. 16, Austria, 1974.

————, "Nomads", *The State of India's Environment, 1982–83: A Citizens' Report,* Centre for Science and Environment, New Delhi, 1982.

Malhotra, K. C., Hulbe, S. K., Khomne, S. B., and Kolte, S. B., "Economic Organisation of a Nomadic Community, the Nandiwallas", *Nomads in India,* (ed.), Misra, P. K., and Malhotra, K. C., Anthropological Survey of India, Calcutta.

Malhotra, K. C., and Khomne, S. B., "Social Stratification and Caste Ranking among the Nandiwallas of Maharashtra", *Proceedings of the Seminar on Nomads in India*, Mysore, 1978.

Malhotra, K. C., Khomne, S. B., and Gadgil, M., "On the Role of Hunting in the Nutrition and Economy of certain Nomadic Population of Maharashtra", *Man in India*, Vol. 63, 1983.

Mani, A., "Social Intervention in the Management of Forest Resources: A Report on the Appiko Movement", *Mimeo*, Indian Institute of Management, Bangalore, 1984.

McEvoy, A. E., "Towards an Interactive Theory of Nature and Culture: Ecology, Production and Cognition in the California Fishing Industry", *The Ends of the Earth: Perspectives on Modern Environmental History*, (ed.), D. Worster, Cambridge University Press, Cambridge, 1989.

Meher Homji, V. M., "History of Vegetation of Peninsular India", *Man and Environment*, Vol. 13, 1989.

Mehra, K. L., and Arora, R. K., "Some Considerations on the Domestication of Plants in India", *Recent Advances in Indo-Pacific History*, (ed.), V. N. Misra and P. Bellwood, Oxford and IBH, New Delhi, 1985.

Misra, V. N., "Ecological Adaptations during Stone-Age in Western and Central India", *Ecological Background of South Asian Prehistory*, South Asia Occasional Papers and Theses, (ed.), Kennedy, K. R. A. and Possehl, G. L., Cornell University Press, Ithaca, 1973.

Mooser, J., "Property and Wood Theft: Agrarian Capitalism and Social Conflict in Rural Society, 1800–1850. A Westphalian Case Study", *Peasants and Lords in Modern Germany*, (ed.), R. G. Moeller, Allen and Unwin, London, 1986.

Moosvi, S., *The Economy of the Mughal Empire c. 1595. A Statistical Study*, Oxford University Press, Delhi, 1987.

Morris, M., *Forest Traders: A Socio-Economic Study of the Hill Pandaram*, The Athlone Press, London, 1982.

Mukherjee, R. K., *The Foundations of Indian Economics*, Longman, Green and Company, London, 1916.

————, *Regional Sociology*, The Century Company, New York, 1926.

Murali A., "Whose Trees", Forest Practices and Local Communities in Andhra, 1600–1922, *Nature, Culture and Imperialism*, (ed.), Arnold, David and Guha, Ramachandra, OUP, Delhi, 1995.

————, "Civil Disobedience Movement in Andhra 1920–22: The Nature of Peasant Protest and the Methods of Congress Political Mobilisation", *Congress and Classes: Nationalism, Workers and Peasants*, (ed.), Kumar, Kapil, Manohar Publications, New Delhi, 1988.

Muranjan, S. W., "Exploitation of Forests through Forest Labour Cooperatives", *Artha Vijnana*, Vol. 16 (2), Gokhale Institute of Politics and Economics, Poona, 1974.

————, "Impact of Some Policies of the Forest Development Corporation on the Working of the Forest Labourers Cooperatives", *Artha Vijnana*, Vol. 22 (4), 1980.

Nair, K. K., "Forest Development in India", *Indian Journal of Agricultural Economics*, Vol. 22, 1967.

National Commission of Agriculture, *Report of the National Commission of Agriculture*, Ministry of Agriculture, Delhi, Vol. 9, 1976.

Socio-Economic Survey of Primitive Tribes in Madhya Pradesh, National Council of Applied Economic Research, New Delhi, 1963.

Nigam, Sanjay, "Disciplining and Policing the Criminals by Birth", *The Indian Economic and Social History Review*, Vol. 27, No. 2, Sage Publication, New Delhi, 1990.

Pandian, M. S. S., "Political Economy of Agrarian Change in Nanchilnadu: The Late Nineteenth Century to 1939", unpublished Ph. D. thesis, University of Madras, 1985.

Parulekar, G., *Adivasis Revolt*, National Book Agency, Calcutta, 1976.

————, *Jevha Manus Jaga Hoto*, (Marathi), Mouz Publications, Bombay, 1965.

Paul, G. P., *Felling Timber in the Himalaya*, Punjab Printing Company, Lahore, 1871.

Pearson, G. F., "Deodar Forests of Jaunsar Bawar", *Selections from the Records of the Government of the Northwestern Provinces,* 2 Series, Allahabad, Vol. 2, 1870.

Peluso, N., "Rich Forests, Poor People and Development", unpublished Ph. D. thesis, Department of Rural Sociology, Cornell University, 1989.

Peoples Union for Democratic Rights, *Undeclared Civil War: A Critique of the Forest Policy,* Peoples Union for Democratic Rights, Delhi, 1982.

Periera, Winin, "The Sustainable Lifestyle of the Varlis in Indegenous Vision: Peoples of India, Attitudes to the Environment of India", *International Centre Quarterly,* Spring Summer, Vol. 19, 1992.

Pimm, S. L., "Community Stability and Structure", (ed.), Soule, M. E., *Conservation Biology: The Science of Scarcity and Diversity,* Sinauer, Sunderland, Massachusetts, 1986.

Pingle, U., Raja Reddy, N. V., and Christoph von Furer Haimendorf, "Should Shifting Cultivation be Banned?", *Science Today,* April, 1982.

Polanyi, K., *The Great Transformation,* Beacon Press, Boston, 1957.

Poffenberger, Mark, McGean, Betsy, Khare, Arvind, "Communities Sustaining India's Forests in the Tewenty-first Century", (ed.), Poffenberger, Mark, and McGean, Betsy, Oxford University Press, Delhi, 1996.

———, *Village Voices, Forest Choices: Joint Forest Management in India,* (ed.) Poffenberger, Mark, and Mcgean, Betsy, OUP, Delhi, 1996.

Possehl, G. L., *Harappan Civilisation: A Contemporary Perspective,* Oxford and IBH, New Delhi, 1982.

Pouchepadass, Jacques, "British Attitude towards Shifting Cultivation in Colonial South India: A Case Study of South Canara District 1800–1920", *Nature, Culture, Imperialism: Essays on the Environmental History of South Asia,* (ed.), Arnold, David, and Guha, Ramachandra, Oxford University Press, Delhi, 1995.

Prabhakar, R., and Gadgil, Madhav, "Maps as Markers of Ecological Change: A Case Study of the Nilgiri Hills of Southern India", *Nature,*

Culture, Imperialism: Essays on the Environmental History of South Asia, (ed.), Arnold, David, and Guha, Ramachandra, Oxford University Press, Delhi, 1995.

Prakash, Om, "The Dutch Factory at Vengurla in the Seventeenth Century", *Medieval Deccan History,* (ed.) Kulkarni, A. R., Nayeem, M. A., and de Souza, T. R., Popular Prakashan, 1996.

Prasad, S. N., and Gadgil, M., "Conservation of Bamboo Resources in Karnataka", *Mimeo,* Karnataka State Council of Science and Technology, Bangalore, 1981.

Preston, Lawrence, W., *The Devs of Cincvad: A Lineage and the State in Maharashtra,* Cambridge University Press in association with Orient Longman, Bombay, 1990.

Prucha, F. P., *The Indians in American Society: From the Revolutionary War to the Present,* University of California Press, Berkeley, 1985.

Rai, S. N., "Protection Forestry with a Reference to Ecological Balance in the Tropical Forests of Western Ghats of Karnataka", National Seminar on Forests and Environment, Karnataka Forest Department, 2–3 December, 1981.

Rana, S., "Vanishing Forests", *Frontier,* Vol. 16 (6), Kolkata, 1983.

Ranade, Mahadev, Govind, "Essays on Indian Economics: A Collection of Essays and Speeches," G. A. Natesan and Company, Madras, 1920.

Randhawa, M. S., "Nature Conservation, National Parks and Bio-Aesthetic Planning in India", Presidential Address, Botany Section, 35 Indian Science Congress, Allahabad, 1949.

Rangarajan, Mahesh, *Fencing the Forest: Conservation and Ecological Change in India's Central Provinces 1860–1914,* Oxford University Press, Delhi, 1996.

Rappaport, R. A., *Pigs for the Ancestors: Ritual in the Ecology of a New Guinea People,* Yale University Press, New Haven, 1984.

Raumolin, J., "Impact of Technological Change on Rural and Regional Forestry in Finland", *Mimeo,* World Institute of Development Economics Research, Helsinki, 1986.

Ribbentrop, Berthold, *Forestry in British India*, Government Press, Calcutta, 1900, reprinted New Delhi, 1989.

Richards, J. F., "World Environmental History and Economic Development", *Sustainable Development of the Biosphere*, (ed.), Clark, W.C., and Munn, R.E., Cambridge University Press, Cambridge, 1986.

Roy, S. L., *The Birhors: A Little Known Jungle Tribe of Chota Nagpur*, K. E. M. Mission Press, Ranchi, 1925.

Ruddle, K., and Johannes, T. E., (ed.), *The Traditional Knowledge and Management of Coastal Systems in Asia and the Pacific*, UNESCO, Indonesia, 1985.

Sagreya, K. P., *Forests and Forestry*, National Book Trust, Delhi, 1979.

Saldhana, C. J., *Andaman, Nicobar and Lakshadweep, An Environmental Impact Assessment*, Oxford and IBH Publishing Company, New Delhi, 1989.

Saldhana, Indra, Munshi., "The Political Ecology of Traditional Farming Practices in Thana District, Maharashtra (India)", *The Journal of Peasant Studies*, No. 3, Vol. 17, 1990, pp. 433–443.

————, "Colonial Forest Regulations", *Nature and Orient: Environmental History of South and Southeast Asia*, (ed.), Grove, R., Damodaran, V., Sangwan, S., OUP, Delhi, 1998.

Sankhala, K. S., and Jackson, P., "People, Trees and Antelopes in the Indian Desert", *Culture and Conservation: The Human Dimensions in Environmental Planning*, (ed.), McNeely, J. A., and Pitt, D., Croom Helm, Dublin, 1985.

Sarkar, S., "Primitive Rebellion and Modern Nationalism: A Note on Forest *Satyagraha* in the Non-Cooperation and Civil Disobedience Movements", *National and Left Movements in India*, (ed.), Panikar, K. N., New Delhi, 1980.

————, *Modern India, 1885–1947*, Macmillan, New Delhi, 1983.

————, "The Conditions and Nature of Subaltern Militancy: Bengal from Swadeshi to Non-cooperation", *Subaltern Studies III*, (ed.), Guha, R., Oxford University Press, New Delhi, 1984.

Savyasachi, "Agriculture and Social Structure: The Hill Maria of Bastar", *Mimeo*, World Institute of Development Economics and Research, Helsinki, January 1986.

Scott, J. C., *The Moral Economy of the Peasant: Rebellion and Subsistence in Southeast Asia*, Yale University Press, New Haven, 1976.

————, *Weapons of the Weak: Everyday Forms of Peasant Resistance*, Yale University Press, New Haven, 1986.

————, "Resistance Without Protest and Without Organisation: Peasant Opposition to the Islamic Zakat and the Christian Tithe", Comparative Studies in Society and History, Vol. 29, 1987.

Sengupta, N., "The Indigenous Irrigation Organisation of South Bihar", *Indian Economics and Social History Review*, Vol. 17, 1980.

————, (ed.), *Jharkhand: Fourth World Dynamics*, Authors' Guild, Delhi, 1982.

Sharma, G. R., *The Beginnings of Agriculture*, Allahabad, 1980.

Sharma, R. S., *Urban Decay in India*, New Delhi, 1987.

Shiva, V, Sharatchandra, H. C., and Bandopadhyay, J., "Social Forestry: No Solution Within the Market Place", *The Ecologist*, Vol. 12., 1982.

Singh, Chetan, "Forests, Pastoralists and Agrarian Society in Mughal India", *Nature, Culture, Imperialism : Essays on the Environmental History of South Asia*, (ed.), Arnold, David and Guha, Ramachandra, Oxford University Press, Delhi, 1995.

Singh, K. S., (ed.), *Tribal Movements in India*, Vol. II, Manohar, New Delhi, 1983.

Sinha, A. C., "Social Frame of Forest History", *Social Science Probings*, Vol. 3(2), 1986.

Skaria, Ajay, "Timber Conservancy, Dessicationism and Scientific Forestry: The Dangs 1840s–1920s", *Nature and The Orient: The Environmental History of South and Southeast Asia*, (ed.), Grove, Richard H., Damodaran, Vinita, and Sangwan, Satpal, Oxford University Press, Delhi, 1998.

Smythies, E. A., *India's Forest Wealth*, Humphrey Milford, London, 1925.

Smythies, A., and Dansey, E. (eds.), *A Report on the Proceedings of the Forest Conference held at Dehradun in October 1886*, Government Press, Simla, 1887.

Sontheimeir, Gunther, Dietz, *Pastoral Dieties in Western India*, translated by Anne Feldhaus, OUP, 1993.

Soule, M. E., (ed.), *Conservation Biology: The Science of Scarcity and Diversity*, Sinauer, Sunderland, Massachusetts, 1986.

Srinivas, M. N., *Caste in Modern India and Other Essays*, Asia Publishing House, Calcutta, 1962.

Stebbing, E. P., *The Forests of India*, John Lane, London, Vol. 1–3, 1922–1927.

Sukumar, R., *The Asian Elephant: Ecology and Management*, Cambridge University Press, Cambridge, 1989.

Surapala, "*Vrikshayurveda: The Science of Plant Life*", (translated by Nalini Sadhle), Asian Agri-History Foundation, Secunderabad, 1994.

Terbourgh, J., "Keystone Plant Resources in the Tropical Forest", *Conservation Biology – The Science of Scarcity and Diversity*, (ed.), Soule, M.E., Sinauer Associates, Sunderland, Massachusetts, 1986.

Tewari, K. M., "Research Needs of Social Forestry", *Indian Forester*, No. 107, 1981.

Thakurdas, P., et al., *Memorandum Outlining a Plan of Economic Development for India*, Penguin, London, 1944.

Thapar, Romila, *A History of India*, Vol. 1, Penguin Books, Harmondsworth, 1966.

——————, *From Lineage to State*, Oxford University Press, Bombay, 1984.

——————, "Significance of Regional History with Reference to the Konkan", *Medieval Deccan History*, (ed.), Kulkarni, K. R., Nayeem, M. A., DeSouza, T. R., Bombay, 1996.

Thompson, E. P., "The Moral Economy of the English Crowd in the Eighteenth Century", *Past and Present*, No. 50, 1970–71.

————, *Whigs and Hunters: The Origin of the Black Act*, Penguin, Harmondsworth, 1975.

Totman, C., *The Origins of Modern Japan's Forests: The Case of Akita*, University of Hawaii Press, Honolulu, 1985.

————, *The Green Archipelago: Forestry in Pre-Industrial Japan*, University of California Press, Berkeley, 1989.

Trautmann, T. R., "Elephants and the Mauryas", *India: History and Thought Essays in Honour of A. L. Basham*, (ed.), Mukherjee, S. N., Subarnarekha, Calcutta, 1982.

Trevor, C. G., Smythies, E. A., *Practical Forest Management*, Government Press, Allahabad, 1923.

Tucker, Richard, P., "Forest Management and Imperial Politics: Thana District", Mumbai, 1823–1887," *The Indian Economic and Social History Review*, No. 3, Vol. 16, 1983.

————, "The Depletion of India's Forests under British Imperialism", *The Ends of Earth: Perspectives on Modern Environmental History*, (ed.), Worster, D., Cambridge University Press, New York, 1988.

Tucker, Richard P., and Richards, J. F., (ed.), *Deforestation and the Nineteenth Century World Economy*, Duke University Press, Durham, 1983.

Upadhyay, C. B., *Forest Laws*, Hind Publishing House, Allahabad, 1984.

A Survey of India's Export Potential of Wood and Wood Products, United States Agency for International Development, Delhi, Vol. 1, 1970.

Vail, D., "Contract Logging, Clear Cuts and Chainsaws", *Mimeo*, World Institute of Development Economic Research, Helsinki, 1987.

Varma, S. J., *Mountstuart Elphinstone in Maharashtra (1801–1827): A Study of the Territories Conquered from the Peshwas*, K. P. Bagchi and Company, Calcutta, 1981.

Vartak, V. D. and Gadgil, Madhav, "Sacred Groves of India: A Plea for Continuous Conservation", *Journal of Bombay Natural History Society*, No. 2, Vol. 2, Bombay, 1975.

Vayda, A. P., "Warfare in Ecological Perspective", *Annual Review of Ecology and Systematics*, Vol. 5, Palo Alto, California, 1974.

Vijayan, V. S., "Keoladeo National Park Ecology Study", *Bombay Natural History Society, Annual Report 1987*, 1987.

Visveswaraya, M., *Reconstructing India*, P. S. King, London, 1920.

Voelcker, J. A., *Report on the Improvement of Indian Agriculture*, Government Press, Calcutta, 1893.

Von Furer Haimendorf, C., *The Chenchus: Jungle Folk of the Deccan*, Macmillan, London, 1943.

———————, *The Reddis of the Bison Hills: A Study in Acculturation*, Macmillan, London, 1943.

———————, *Tribal Hyderabad: Four Reports*, Government Press, Hyderabad, 1945.

———————, "Aboriginal Rebellions in the Deccan," *Man in India*, No. 25, 1945.

———————, *Tribes of India: The Struggle for Survival*, Oxford University Press, Delhi, 1985.

Von Munroy, J. A., *Report to the Government of India on Integration of Forests and Forest Industries*, Food and Agricultural Organisation, Rome, 1960.

Von Tunzelman, G. N., "Technical Progress during the Industrial Revolution", *The Economic History of Britain Since 1700*, Floud, R. (ed.), and McCloskey, D., Cambridge University Press, Cambridge, Vol. 1: 1700–1860, 1981.

Wade, R., *Village Republics: Economic Conditions for Collective Action in South India*, Cambridge University Press, Cambridge, 1988.

Wallerstein, I., *The Modern World System*, Academic Press, New York, 1974.

Walton, H. G., *British Garhwal: A Gazetteer*, Government Press, Allahabad, 1910.

Ward, H. C., *Report of the Land Revenue Settlement of the Mundlah District of the Central Provinces (1868–69)*, Government Press, Bombay, 1870.

Webb, W. P., *The Great Frontier*, University of Texas Press, Austin, 1964.

Webber, T. W., *The Forests of Upper India and Their Inhabitants*, Edward Arnold, London, 1902.

Whitcombe, E., *Agrarian Conditions in Northern India, Volume I: The United Provinces Under British Rule, 1860–1900*, University of California Press, Berkeley, 1971.

White, L., "The Historical Roots of our Ecological Crisis", *Science* No. 155, 1203–5, American Association for the Advancement of Science, Washington D.C., 1967.

Wilkinson, R. G., "The English Industrial Revolution", Worster, D., *The Ends of the Earth : Perspectives on Modern Environmental History*, Cambridge University Press, Cambridge, 1989.

Wink, Andre., *Land and Sovereignty in India: Agrarian Society and Politics under the Eighteenth Century Maharashtra Swarajya*, Cambridge University Press, 1986.

World Commission on Environment and Development, *Our Common Future*, Oxford University Press, New Delhi, 1987.

Worster, D., (ed.), *The Ends of the Earth, Perspectives on Modern Environmental History*, Cambridge University Press, Cambridge, 1988.

Zimmerman, *The Jungle and the Aroma of Meats: Ecological Themes in Hindu Medicine*, London, 1989.

Index